Making Measures

Benjamin D. Wright and Mark H. Stone

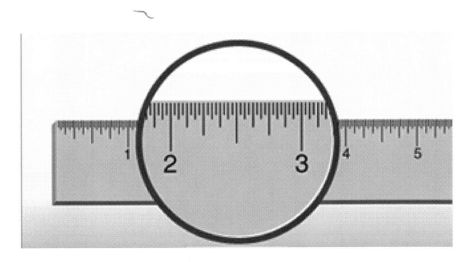

The Phaneron Press • Chicago

The Phaneron Press, Inc.
1252 W. Catalpa Ave.
Chicago, IL 60640-1337
mail@thephaneronpress.com

Manufactured in the United States of America.

Benjamin D. Wright and
Stone, Mark H.
Making Meausres.
    Includes bibliographical references.
    Includes index.
    Content: 1. Measurement—probablity, Rasch measurement,
fit statistics. 2. Science—observation, constructs. 3.
Psychometrics—Knox cube test, reading ability, readability.
    ISBN 0-9749871-1-5 (paper)
    Library of Congress Control Number: 2003109922

Contents

Figures

Tables

Introduction

This book explains, step by step, how to construct measures from observations. It outlines a method for collecting, sorting, organizing, and mapping data to facilitate telling a story about what the data mean.

The data can be generated by aspects of our environment (the world around) or gleaned from aspects of our thoughts, feelings, responses (the world inside). The methodology is based on well-founded, centuries-old principles and procedures of mathematics and measurement.

Chapters 1 through 6 build the foundations of measurement. Chapters 7 and 8 show how to apply these age-old principles using software based on the Rasch measurement model.

Chapter 9 is an adaptation of an Association of Test Publishers Presentation by A. Jackson Stenner with Ben Wright, which was held in honor of Ben in San Diego, California (February, 2002).

Foreword

Ben Wright and Mark Stone's insights into the philosophical and mathematical foundations of measurement are necessary for constructing practical, objective assessments of educational, psychological, and physical functioning. Since the publication of their seminal text, *Best Test Design* (1979), they have collaborated on a number of works designed to make accessible the principles and procedures of objective measurement. This volume widens yet again the audience that can benefit from these ideas. By beginning with a disarmingly simple stone-lifting problem the reader is introduced step by eloquent step to the thinking and mathematics behind making measures and making measures meaningful. This volume looks back at the philosophical underpinnings of measurement and connects the prescience of Charles S. Peirce (1838–1914) to the models of Georg Rasch (1901–1980) and finally to the day-to-day needs of instrument-builders in the human sciences.

From the perspective of *Best Test Design*, this volume also moves forward with an emphasis on the role of substantive theory in measurement. The measurement applications, the Knox Cube Test for attention and short term memory (Chapter 7), Wolpe and Lang's (1964) Fear Survey Schedule (Chapter 8), and the Lexile Framework for Reading (Chapter 9), make extensive use of theory in predicting why item difficulties vary. Can we imagine a more important piece of evidence for the construct validity of an instrument than a specification equation that is capable of explaining variation in observed item difficulties?

The paradox of unity and separation finds expression in the puzzle of whether a book is well comprehended by a reader because the book is easy or because the reader is skilled. Does a person succeed on an attention task because it requires fewer memory registers or because the person possesses good attention? Is a stone lifted because the stone is light or the lifter is strong? This paradox presents in various guises across the human sciences or more generally wherever measurement is contemplated. The paradox is resolved by positing a single yardstick of reading ability/readability, attention/difficulty, or strength/weight. This book explains how and why such yardsticks are built and how to ensure quality in their construction.

A. Jackson (Jack) Stenner
MetaMetrics CEO
Durham, North Carolina
April 21, 2003

Chapter 1 *Successful Science*

Successful science depends on the intuition, explanation and application of useful ideas. As a promising idea takes shape, we set out to make observations that are sufficiently focused on its implications to produce useful evidence of its utility. Science is an evolving dialogue between idea and experience, theory and observation. Theory guides the way we track and organize our observations. Appreciation of what we observe improves the articulation and utility of our theory. Theory and observation interact dynamically to produce better science and better living. But observations cannot be random. Unless our observations develop a direction, they remain empty.

Consider beach-combing. If we gather shells at random, what might we observe? What guides us as we make repeated observations? Even *same* and *different* require context. Shells might be considered all the same when compared to stones. But shells can differ from each other in many ways: size, color, shape. To progress, we have to decide what we are seeking. Our observations must track with our ideas. Our ideas must weigh our observations and convey them into relevant data.

Chapter 2   *Sameness*

Productive shell-collecting depends on introducing a speci-
fied sameness into what we observe.

　　Two shells are either the same or different on the
attribute by which we decide to compare them. If they are
different, our "sameness" odds remain an even 1 to 1. The
evidence for defining what is "same" and hence what is "different" remains equivocal. We need more than two shells.

　　How many shells do we need to establish a believable
sameness? Add another shell. Now we either have two of one
type and one of another, or else we have three disparate
instances. No definition of sameness can be established
without the observation of something different. No sameness
can be constructed without a decision as to what to call
"same." But no sameness can be pursued without at least
two examples of what we mean by "same." If two of three
shells can be said to be similar and one different, then our
sameness odds are 2 to 1. We need three instances to establish a 2-to-1 definition of sameness. Further instances of
sameness increase our odds and hence our confidence in the
utility of what we are thinking and doing. More replications
of sameness of an attribute make it more established and
more able to expose instances of difference.

The principle that guides the evaluation of observations is *sameness replication*. We track to accumulate more and more instances of what we are learning to name as examples of a specific sameness. Different begins as "not the same." It has no meaning of its own until we accumulate enough "same" observations to make clear what something "different" must be. Before that point is reached, everything is different from everything else, and so "different" remains meaningless. Sameness implies regularity in what we are observing. Contradictions to this regularity are meaningful only to the degree that we have established a clear definition of sameness. It takes replications of sameness to build a position and so to bring out what then become instances of difference. Tracking our observations to increase the odds for sameness is the only way to build confidence in what we are seeking.

Chapter 3   *Conjunction*

We see a man lift a stone. How do we understand this raw observation? Does the stone rise because the stone is light or because the man is strong?

The unavoidable conjunction of man and stone in this observation provokes a question true of all raw observations. Every observation is evidence of a conjunction of at least two forces. Until we invent a way to make separate estimates of the forces involved, any conclusion we might wish to reach with respect to the meaning of the observation is confounded by this conjunction.

How strong is the man?

How heavy is the stone?

To answer these questions we must replicate our observations. We must bring together several men and several stones, ask the men to attempt to lift the stones and record which men lift which stones. This simple data matrix of $X_{ni} = 1$, when man n lifts stone i, and $X_{ni} = 0$, when he does not, can lead to a basis for comparing men on their strength and stones on their weight, independently of one another.

If we can position these seemingly different but in fact conjoint variables, strength and weight, on a single strength/weight yardstick, we will be able to make inferences about all possible comparisons among forces exerted by men and masses manifested by stones. We will be able to mark out or to calibrate a strength/weight yardstick from the results of pitting man-strength B_n against stone-weight D_i.

Charles Sanders Peirce (1878) wrote,

> It is incontestable that the chance for an event has an intimate connection with the degree of our belief in it. Any quantity which arises with a chance might, therefore, serve as a thermometer for the intensity of belief. When there is a very great chance, the feeling of belief is very intense. As the chance diminishes, the feeling of believing should diminish, until an even chance is reached, where it should completely vanish. When the chance becomes less than even, then a contrary belief should spring up and should increase in intensity as the chance diminishes, and as the chance almost vanishes the contrary belief should tend toward an infinite intensity. Now, there is one quantity which, more simply than any other, fulfills these conditions; it is the logarithm of the chance (log odds). But there is another consideration which must fix us to this choice for our thermometer. It is that our belief ought to be proportional to the weight of evidence. Two arguments which are entirely independent, neither weakening nor strengthening each other, ought, when they concur, to produce a belief equal to the sum of the intensities of belief which either would produce separately. The chance of independent concurrent arguments are multiplied together to get the chance of their combination. But the quantities which best express the intensities of belief should be such that they are to be added when the chances are multiplied in order to produce the quantity which corresponds to the combined chance. The logarithm of the chance is the only quantity which fulfills this condition.

The rule for the combination of independent concurrent arguments takes a very simple form when expressed in terms of the intensity of belief, measured in the proposed way. Take the sum of all the feelings of belief which would be produced separately by all the arguments pro, subtract from that the similar sum for arguments con, and the remainder is the feeling of belief which we ought to have on the whole. These considerations constitute an argument that the conjoint probability of all the arguments in our possession, with reference to any fact, must be intimately connected with the just degree of our belief in that fact. (p. 709)

It follows that only a conjoint additive specification such as $(B_n - D_i)$ that is directed to govern the conjunction of a man lifting a stone can enable us to construct a useful strength/weight yardstick that separates the man-strength and stone-weight forces that are conjoined in our observations. An observable intersection of strength and weight can be pictured (Figure 1) as

$$\text{Stone-weight} \quad D_i$$

$$\text{Man-strength } B_n \text{ —— } X_{ni} = \text{either } 0,1 \quad (1)$$

where

"0" indicates that stone-weight overcomes man-strength and "1" indicates that man-strength overcomes stone-weight.

Figure 1. *Observable Intersection*

When man n is stronger than stone i is heavy, then B_n is greater than D_i, their difference $(B_n - D_i)$ is greater than 0, and the probability that man n lifts stone i, P_{xni}, is

greater than ½. But when man n is weaker than stone i is
heavy, then B_n is less than D_i, their difference $(B_n - D_i)$ is
less than 0 and the probability, P_{xni}, that man n lifts stone i
is less than ½.

Thus

$$B_n > D_i \text{ makes } (B_n - D_i) > 0 \text{ so } P_{xni} > .5 \quad (2)$$

$$B_n < D_i \text{ makes } (B_n - D_i) < 0 \text{ so } P_{xni} < .5 \quad (3)$$

From which follows

$$B_n = D_i \text{ makes } (B_n - D_i) = 0 \text{ and } P_{xni} = .5 \quad (4)$$

To study conjunctions of strength and weight we can
collect a matrix of man/stone comparisons produced by men
who differ in strength attempting stones that differ in
weight. From these data we can build a single dimension
upon which men and stones are located and from which their
future behavior can be predicted.

Far more important than the already out-of-date fact of
this data matrix are its predictive possibilities. Only successful prediction makes experience useful. Our analysis of
prior man/stone observations can enable us to estimate the
outcomes of future man/stone conjunctions, to predict what
will happen next. We can build and sharpen these predictions by enriching our experience with more replications,
more observations of men lifting stones. These observations
are designed to track and strengthen the construction of our
yardstick for measurement of a strength/weight variable.

Picture our observations as a data matrix of comparisons made from conjunctions of five men and six stones
arranged by strength and weight and, for the moment,
showing no surprises and no missing data (Figure 2).

		Light		Stones		Heavy		Men Scores
		1	2	3	4	5	6	
Weak	1	1	0	0	0	0	0	1
	2	1	1	0	0	0	0	2
Men	3	1	1	1	0	0	0	3
	4	1	1	1	1	0	0	4
Strong	5	1	1	1	1	1	0	5
Stone Scores		5	4	3	2	1	0	

Figure 2. *"Idealized" Guttman Diagram*

This idealized outcome conforms to a *Guttman diagram* (Guttman, 1944) of conjoint transitivity because
stones, men, and their conjunctions are perfectly ordered.
The triangle of successful responses, evident in the lower
left of the matrix, expresses the steadily increasing strength
of men overcoming the steadily increasing weight of stones;
similarly, the triangle of failures expresses the steadily
increasing weight of the stones overcoming the men's
strength. This perfect outcome makes our ideals clear. But it
is too perfect to observe.

A more probable and hence more useful outcome is to
imagine instead that what are perfectly ordered is not our
observations but rather our ideas of what is producing them.

A perfect Guttman ordering of predictors is the canonical definition of a useful measurement system. The observed values, however, subject as they are to the vicissitudes of reality, can only be stochastic — probable consequences of their perfectly ordered predictors. Each outcome remains a single event, $X_{ni} = 0$ or 1, but is understood as the result of a probability, $0 < P_{xni} < 1$, rather than as a certainty — a probability that specifies a Bernoulli (1713) binomial outcome $X_{ni} = 0$ or 1. The P_{xni} are perfectly ordered by our idea of the additive conjoint specification $(B_n - D_i)$, which we define and require them to follow. The stochastically disordered raw observations are in keeping with what we experience in real life: sometimes things occur as expected, sometimes not.

A corresponding matrix of real observations is shown in Figure 3.

		Light		Stones		Heavy		Men Scores
		1	2	3	4	5	6	
Weak	1	1	0	0	0	0	0	1
	2	1	0	0	0	0	0	1
Men	3	1	1	1	0	0	0	3
	4	1	1	1	0	1	0	4
Strong	5	1	1	1	1	1	0	5
Stone Scores		5	3	3	1	2	0	

Figure 3. *"Realistic" Results Guttman Diagram*

All that remains is to connect the outcome probabilities to a men/stones, strength/weight dimension. The connection must follow the form of a monotonically increasing ogive (Figure 4).

Probability P_{1ni} of $X_{ni} = 1$

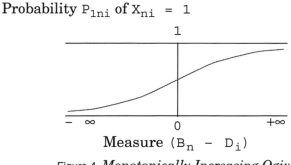

$$Measure\ (B_n - D_i)$$

Figure 4. *Monotonically Increasing Ogive*

The horizontal axis is the dimension to be measured. The vertical axis is the probability of observing evidence of the dimension in a positive response, $X_{ni} = 1$.

Chapter 4   *Men / Stones –*
Strength / Weight

The observable and hence bounded outcome of probabilities $0 < P_{xni} < 1$ that man n will be observed to lift stone i to make $X_{ni} = 1$ must be connected to the boundless and unobservable dimension defined by the additive comparison of man and stone, $(B_n - D_i)$, in a useful way. As Peirce pointed out, the continuously rising ogive is linearized by taking the logarithm, $\log(P_{1ni}/P_{0ni})$, of the success odds, (P_{1ni}/P_{0ni}). This "straightens" the curve and defines $(B_n - D_i)$ as

$$(B_n - D_i) = \log(P_{1ni}/P_{0ni}) \qquad (5)$$

The resulting definition of measurement is called the Rasch model for dichotomous data because the Danish mathematician, Sir Georg Rasch (1961, 1960/1980) was the first person to apply it to constructing measures.

We observe men lifting stones in order to accumulate replications of our idea about a strength/weight yardstick. We record failures and successes, $X_{ni} = 0$ or 1, understand these events as the results of a Bernoulli binomial process governed by $(B_n - D_i)$, and use them to estimate P_{xni} and thence B_n and D_i. We fit the observed data matrix of X_{ni}s to Rasch's model by calculating the estimates of B_n and D_i, which minimizes the difference between the P_{Xni} and the raw experience, X_{ni}. We model a raw score performance as the sum of these modelled probabilities of a correct response over a collection of encounters between a man and a set of stones.

When that is done, we evaluate our success by comparing each observation $X_{ni} = 0$, when man n fails to lift stone i, and $X_{ni} = 1$ when man n succeeds with his expectation, P_{Xni}, to find out how well these data help us to construct a yardstick that measures the ideal abstract forces of strength and weight that manifest in the observable, concrete behavior of men lifting stones. The lifting of stones is a fact from experience. The implied forces are a fiction. But it is the reproducibility of the fiction that gives the facts meaning and value.

Chapter 5   Newton's Second Law of Motion

Suppose, now, we define the force of man-strength B_n as

$$F_n = \exp(B_n), \qquad (6)$$

the mass of stone-weight D_i as

$$M_i = \exp(D_i), \qquad (7)$$

and the odds for an observation of the acceleration that occurs when man n causes stone i to leave the ground as

$$A_{ni} = P_{1ni}/P_{0ni}. \qquad (8)$$

Then the exponential form of the Rasch/Peirce model,

$$(P_{1ni}/P_{0ni}) = \exp(B_n)/\exp(D_i) \qquad (9)$$

becomes

$$A_{ni} = F_n/M_i \qquad (10)$$

$$\text{i.e., } F_n = M_i A_{ni}, \qquad (11)$$

and we discover Newton's Second Law of Motion in our conjunction of a gang of men and a field of stones. The evidence was there all the time, thousands of years before Newton wrote down his formulation.

With estimates of the strength B_n of man n and the weight D_i of stone i, we can use their conjunction $(B_n - D_i)$ to calculate the P_{Xni}, which our measurements predict for outcome X_{ni}, where $X_{ni} = 0,1$ and

$$P_{Xni} = \exp(B_n - D_i)/[1+\exp(B_n - D_i)], \qquad (12)$$

and also the Bernoulli variance, $[P_{Xni}(1 - P_{Xni})]$, which quantifies the extent to which we expect instances of this comparison $(X_{ni} - P_{Xni})$ to vary, when our data fit the measurement model.

This enables us to compare observation X_{ni} with its expectation P_{Xni}, calculate the score residual, $(X_{ni} - P_{Xni})$, and scale it by root, $[P_{Xni}(1 - P_{Xni})]$, to give it an expected mean of zero and standard deviation of one. The resulting standardized discrepancy is

$$Z_{Xni} = (X_{ni} - P_{Xni})/[P_{Xni}(1 - P_{Xni})]^{\frac{1}{2}}. \qquad (13)$$

Notice that

$$Z_x^2 = (X - P)^2/[P(1-P)] \qquad (14)$$

so that

$Z_0^2 = P/(1 - P) =$ odds against $X = 0$ (i.e., against failure to lift), and

$Z_1^2 = (1 - P)/P =$ odds against $X = 1$ (i.e., against success to lift).

When the absolute value of z_{Xni} is less than 2, the odds against the observation X_{ni} are less than $z^2 = 4$ to 1. We could accept such an X_{ni} as a not-too-unreasonable consequence of its governing parameters, $(B_n - D_i)$. But when the absolute value of the standardized discrepancy z_{xni} between observation X_{ni} and expectation P_{xni} is more than 3, so that the odds against this X_{ni} being no more than a random aberration have risen to $z^2 = 9$ to 1, we may begin to doubt the fit of this particular X_{ni} to the yardstick we are building. At this point we may not be willing to accept X_{ni} as a useful observation and will investigate its source and diagnostic implications.

We compute the discrepancy z_{Xni} for every instance of X_{ni}. To evaluate a total level of discrepancy for a set (for example, all data from a particular man or stone), we can average the squares of these discrepancies z_{Xni} over any set of men or stones. This brings to our attention whatever inconsistencies lurk in our data.

We began by asking how we might make use of the raw experience of a man lifting a stone. We described how to use observations of men lifting stones to construct a single yardstick to measure a strength/weight, force/mass variable and discovered how this revealed Newton's second law of motion. Finally, we showed how our expectations could be used to judge whether or not any observation or collection of observations is meeting our expectations and hence might be helpful for predicting what will probably happen next.

Chapter 6   Judging Misfit

How shall we evaluate exceptions, that is, the "misfits" encountered? We can do this at the quantitative (statistical) level and at the qualitative (person/item content) level.

Quantitative Level

There are three approaches to evaluating exceptions at the quantitative level: Principal components of response residuals, individual response residuals z and z^2, and summaries of z^2.

Principal Components of Response Residuals. This analysis evaluates the response residual similarities among items and among persons to identify clusters of items or persons with similar patterns among what we would like to regard as nothing more than random, hence meaningless, residuals. Principal component analysis of response residuals among items reveals the presence of unsuspected secondary variables contained in item content. A frequent example is a subset of negatively worded items that have been reverse

scored in the hope that this will align them with the positive
items — a psychologically naïve maneuver that seldom
works in the manner presumed. Principal component
analysis of response residuals among persons brings out the
presence of subgroups of persons with similar response bias.
Frequent sources are gender, first language, and ethnicity.

Principal component analysis of stone residuals might
identify a subset of stones that has something in common.
When we examine these stones, we might find that smooth
stones are harder to lift than rough stones of similar weight,
which would produce a tell-tale set of similar residuals. This
would identify a secondary and probably unwanted variable
of smoothness operating in our men/stones data and give us
the opportunity to decide whether or not we want to measures stones on two variables (i.e., weight and smoothness)
or control the intrusion of smoothness by ensuring that all of
the stones that we use to build our strength/weight measures are equally smooth. When constructing a strength/
weight yardstick we would then take care to use stones of
similar smoothness in order to clarify our definition of
strength/weight. Principal component analysis of men residuals might also show a second variable — this time the effect
of wet hands on lifting. The natural resolution of this disturbance to the construction of a strength/weight yardstick is to
control for hand wetness.

Principal component analysis exposes the presence and
sources of any differential item or person that is functioning
actively in the data. If no salient components are found, then

we know that there is no evidence of person bias or differential item functioning (DIF) in these data. There is no better
or simpler way to detect, identify, and control systematic
bias and differential item functioning than the information
provided by principal component analyses of response
residuals.

Individual Response Residuals: Z and Z^2. We can calculate
the discrepancy between what our measurement system
expects and what has been observed, square the difference,
divide it by its expected variance, and calculate the odds
against that observation occurring by chance. A Guttman
diagram that shows the observed value of every unexpected
X_{ni} (when absolute $Z_{xni} > 2$) at its row n and column i
enables us to see immediately which persons and which
items are producing improbable X_{ni}s. Table 1 on page 22
shows inconsistent responses in the data matrix (i.e., 1s in a
pattern of 0s and 0s in a pattern of 1s).

Table 1. *Most Unexpected Response Guttman Diagram*

PUPIL	MEASURE	ACT
		1111112 122 1 22
		89203112542669784035
		high-------------------
41 FXXXXX, NATASHA	4.77	................1.
17 SXXXXXXXX, GAIL	3.55	0
71 SXXXXXX, DAVE	.96 B	.0.10....0.......222
53 SXXXX, ANDREW	-1.59 L	..............1....1
		---------------low
		11111122122619784225
		8920311 542 6 03

Note: WINSTEPS Tables 6.5 and 10.5 v3.08 output, in which "Unexpected Observations [dia-grams] display the unexpected responses in Guttman scalogram format. The Guttman Scalogram of unexpected responses shows the persons and items with the most unexpected data points (those with the largest standardized residuals) arranged by measure, such that the high value observations are expected in the top left of the data matrix, near the 'high,' and the low values are expected in the bottom of the matrix, near the 'low.' The category values of unexpected observations are shown. Expected values (with standardized residu-als less than $|2|$) are shown by '.'. Missing values, if any, are left blank" (Linacre, 2003, p. 174).

Summaries of Z^2. When we have many items and persons, the study of individual response residuals can become over-whelming. Summaries of the response residuals of each item and each person help us to locate item- and person-based sources of disturbance. Two summaries are calculated, outfit and infit.

An outfit mean square residual U is the average Z_{ni}^2 over any selected set of responses, usually person responses to some item i or items responded to by some person n:

$$U_i = \Sigma_n (Z_{ni}^2) / \Sigma_n 1 \qquad (15)$$

summed over all n responding to item i,

$$U_n = \Sigma_i (Z_{ni}{}^2) / \Sigma_i 1 \qquad (16)$$

summed over all i responded to by person n.

An infit mean square residual V differs from an outfit in that each $Z_{ni}{}^2$ is weighted by its information potential V_{ni} of response X_{ni}, which, when $X_n i = 0, 1$ is $[P_{xni}(1 - P_{ni})]$:

$$V_i = \Sigma_n (V_{ni} Z_{ni}{}^2) / \Sigma_n V_{ni} = \Sigma_n (X_{ni} - P_{ni})^2 / \Sigma_n V_{ni} \qquad (17)$$

summed over all n responding to item i,

$$V_n = \Sigma_i (V_{ni} Z_{ni}{}^2) / \Sigma_i V_{ni} = \Sigma_i (X_{ni} - P_{ni})^2 / \Sigma_i V_{ni} \qquad (18)$$

summed over all i responded to by person n.

The outfit mean square residual U is sensitive to off-target (i.e., markedly unexpected) responses, as when a man who is unable to lift more than a few stones unexpectedly lifts a stone that few men have been able to lift. The infit mean square residual focuses instead on the on-target responses that carry the most potential information, as when men work on stones with weights near their strength limits.

Qualitative Level

Theory to Practice. Prior to analysis, our preliminary ideas about the items and persons we choose to study obligates us to form specific hypotheses about both items and persons. Every useful investigation is guided by explicit hypotheses. To maintain a continuous relationship between our theory and our analysis of the data, it is helpful to code the indicators of these hypotheses into our item and person labels. These labels perform a vital function, but only work when careful thought is given to their coding prior to analysis.

Good theory includes explicit hypotheses about item and person hierarchy, specifies an expected difficulty order among items, and spells out the reasons for this specification. We might expect large stones to be heavier than small ones. We want to anticipate what measure order among persons can be explored: How and why do we expect individuals to differ in strength? Do we expect large men and young men to be stronger than small men or old men?

Persons should be labeled by whatever person categories guide the investigation (e.g., age, gender, ethnicity, diagnosis, treatment, first language). Include in the person labels indicators for every person characteristic that is hypothesized to matter. Item labels should contain a clear verbal abstract of item text and should indicate item type and format. Replication of items written to work together should be seen to point in the same direction in the analysis. The use of explicit item labels enables us to see immediately in the analysis output whether our hypothesized intentions actually occur in our data. Reversing the scoring order of negatively worded items to aim all responses in the same direction, requires that these items be labelled as reversed. Do not assume that a simple reversal will suffice. It is usually observed that reversed "negative" items indicate something different than their unreversed "positive" counterparts. Always check to see whether reversed items actually fit with their unreversed equivalents. The absence of fit indicates that two quite different variables have been evoked and detected.

Type of response format should also be coded into each item label. Some items may invite dichotomous responses, others polytomous. Code your item labels accordingly. The same can be done with other differences in response format. Variations in rating formats, such as items scored 1, 2, 3, vs. 1, 2, 3, 4, need to be indicated in the item labels so that analyses are contingent upon response format as well as content.

Once we have explicit person and item labels, we are ready to explore the content of your observations. At the beginning of each analysis, it is important to reflect upon the purpose and hypotheses of the investigation. Analysis must complement intent. We bring together our observations and the analyses we propose to conduct. This continues the dialogue that we began at the outset of our investigation. Analysis is aimless unless we guide it with intention and examine whether or not there is a useful collaboration between our intentions and our results. The hypotheses that prompted the investigation need to be reviewed continuously as outcomes arise so that, as we navigate the analyses, it remains clear where our attention should be directed.

Examining Data. Always have a copy of your original data-gathering instrument (e.g., questionnaire) in hand when proceeding with the analysis. Mark the instrument with the codes that are entered into the item labels, including reversed items, response formats, construct topics. This helps everyone reviewing the data to check continuously for the consistencies and inconsistencies between instrument intentions and observed results.

A useful strategy is to make a preliminary run of 20–30 cases to check the utility of the analysis control file and data labeling. Time and trouble can be saved by using a preliminary run to assure that everything is operating as intended and that no unresolved matters or oversights are evident before beginning the full analysis. Then the full data set can be run without the inconvenience and delay of unforeseen misadventures that could have been corrected earlier.

Item Polarity. Careless data analysis produces embarrassing consequences. The essential first step is to verify the coherence of the data. Item polarity analysis checks whether items have been keyed as intended, whether there are problems in data coding, and, with rating scales, whether the continuum gradients intended among the rating categories has occurred. Such an elementary step might seem unnecessary, but this step is often essential.

Two statistics are useful: item response by measure correlations and infit mean squares. A positive item by measure correlation indicates that the coding of that item is working in the right direction. When negative correlations are observed, it is necessary to return to the item text and rating scale to find out what has produced this unexpected consequence. The choices, at this point, are to fix the problem by rescoring the rating scale or to omit the item. When the item measure correlations are all positive, a useful rule of thumb is to examine items with infit means squares > 1.5 and then to diagnose the reasons for their occurrence. If severe infit misfit is found in only a few items and no useful explanation emerges, omission of these items is the simplest solution. But do not forget the omitted items, which were included because they were hypothesized to fit. Their unexpected misfit is worth reconsideration and diagnosis.

Rating Scale Structure. The mean measures for the responses in each rating scale category should increase as the categories step up the scale in the direction defined as "more." When category mean measures do not advance or are so close in average category measure that they fail to articulate the categories, we can usually improve the information efficiency of our yardstick by combining these adjacent categories. Also consider combining with their nearest neighbor any categories that manifest substantial misfit. But do not forget, when following the evolving numbers, that we must, in the end, be able to explain what we have done in terms of our initial hypotheses and intentions.

Examine responses to each item separately to see whether or not each item is showing the same increasing behavior for its rating categories 0, 1, 2, 3, and so on along the reach of the yardstick. Careful thought is required to fix a rating scale that is not operating as planned. The fix involves rethinking the initial choice and labeling of categories, investigating the frequency of category use, combining adjacent categories when indicated, and going back and forth between scale and output to identify, understand, and repair the response category problems encountered.

Be sure to enter category names in the control files so that they will appear on each category table to assist in identifying and diagnosing problems. Consider combining categories which at first seem incongruous. If most respondents choose STRONGLY AGREE or STRONGLY DISAGREE, the two middle categories AGREE and DISAGREE often signify a similar resistance to making a clear choice and thus can be usefully combined into one "hesitation" category.

Pivot Anchoring. Whenever we combine more than one kind of rating scale on the same yardstick, we need to study how the successive rating categories of each kind of item align best with the implied hierarchy of the yardstick. The pivot point that clarifies this alignment could be at any category except the first. This consideration has nothing to do with fit or measure: its only effect is upon the item hierarchy printed on the item map. The pivot point specifies the category at which X_{ni} would be scored 1 were that item dichotomized.

Monitoring Results. The aim is to construct a scale that serves the intent. If this is not evident in the output, something must be done. Scales are not completed by intent alone. They require an evolving dialogue between intentions operationalized and evidence gathered. Resist the impulse to interpret results before taking ample time to determine whether the scale is producing results that are worth interpreting! Keep an eye on the person and item error-corrected standard deviations and separation indices to monitor progress. Useful improvements of the scale will produce an increase in item and/or person error-corrected standard deviations and an increase in their separation statistics.

At this juncture, we can postpone delving into the secrets of each errant item. Investigate whether there is a general structure among most of the items. We need an overview. If only an item or two are in trouble, the diagnosis of their disposition can be left until later. Changes in a few items at this point will not improve the measures.

Checking Item Dimensionality. A principal components analysis of response residual similarities among items reveals whether there are clusters of items that suggest the presence of an unexpected *second component* in the data. Are there residual clusters of items? If so, more than one dimension may be active. If not, the dimensional claim of the yardstick under construction is confirmed. When no secondary factors among residuals appear, we can conclude that there is no evidence of differential item functioning (DIF) in the data.

A substantial cluster of items on the first factor usually indicates a *branch* of items that are distinguishable from the main stem. This occurs whenever the set of items consists of a main component with a minor subdivision. Social science examples are the mental vs. physical aspects of well-being and the external vs. internal aspects of self-awareness. The identification of subdimensions need not be a problem. But the underlying structure needs to be clarified in order for the analysis to make sense, for the data to be understood, and for the yardstick to become useful.

The variance magnitude of the yardstick evaluates factor strength. Compare the factor variance with the yardstick variance to see whether the ratio is substantial. When the factor variance is only a small part of the yardstick variance, the factor variance need not be given much consideration vis-a-vis stable measurement. When the factor variance is large compared with the yardstick variance, then its influence must be clarified and implemented, usually as a second dimension leading to two different yardsticks.

A principal components analysis of response residual similarities among persons exposes whatever gender, age, or ethnicity issues bear on instrument bias and hence also on differential item functioning. When person labels have been used to identify categories of persons and indicators of research hypotheses, there is a rich reward from this kind of analysis.

Refining the Yardstick. Now it is time to refine the yardstick, to go back and fine-tune. Items are the first focus of our attention in the construction of a yardstick for measuring a variable. Reconsider the most misfitting items. If they do not enrich the definition of the yardstick and are not needed for greater precision, remove them. Try removing errant items one at a time until person separation begins to decrease. The usual approach is to start with the most misfitting items. Check item and person error-corrected standard deviations and separations at each step to see whether item removals produce improvement.

Pay attention to the measured spacing between items. Items should spread evenly across the intended range of the instrument. Work toward having multiple items at a given location when close decisions, such as "pass/fail," occur at that location. Enough items should obtain measures that are sufficiently accurate for the purpose at hand, but no more. There is no reason to use more items than produce the desired level of efficiency. Do not, however, remove substantively crucial items solely on the basis of misfit. When an item text contains meaning that is essential to the intent, think twice before discarding it for statistical reasons. Item removal requires dialogue between the subject of inquiry and the obtained responses. Each item deletion should have a rational basis.

Checking Persons. The items that define the yardstick should provide the generality we seek. We also need to be aware that the persons are local and transient. Our goal, however, is to build a yardstick for measuring everyone yet to be evaluated, not just those residuals whose data are at hand. We want the sample of persons in the study to be typical of the whole population of persons for whom the yardstick is intended.

After careful consideration of items, address the persons. Investigate the measures of persons who are expected to measure high and low. List persons according to their degree of misfit. In general, because the goal is to build a yardstick, first attention has been given to the items. At this point it is reasonable to expect the yardstick to endure. Samples are always local and always suspect. Nevertheless, the sample has been designed to include relevant persons. Which of these persons threaten the yardstick? Which persons appear mismeasured? Check misfits against their person labels. Use Guttman patterns to identify idiosyncratic persons. Reach for an understanding of what went wrong for the persons who misfit.

Fit statistics, which describe the immediate relation between intentions and data, are as local and transient as the data. Statistics must be transcended by a clear understanding of the construct implied. This requires attention to the content of the items and their influence on the fit statistics. Fit statistics alone cannot provide all the information that is needed to make good decisions about building yardsticks for measuring. Knowledge of item content and of the nature of the persons involved are essential to understanding how to use misfit to advantage.

Chapter 7   *Knox Cube Test–Revised*

This chapter and the next illustrate constructing a measure with WINSTEPS, a versatile and comprehensive Rasch measurement software program (Linacre, 2003).

Constructing a Measure

The *Knox Cube Test–Revised* (KCT–R) (Stone, 2002) data are 26 tapping patterns administered to 2161 tested clients of a metropolitan outpatient clinic. Figure 5 on page 36, a plot generated with the measurement software, is a *key map* of the relationship between the difficulty of the 26 tapping patterns (located on the right side of the vertical axis in difficulty order from the easiest at the bottom to hardest at the top) and the 2143 analyzed person measures distributed along the horizontal axis (in ability order from least able on the left to most able on the right). Tapping patterns are identified by their administration order (and theoretical difficulty order), labeled NUM, and their tapping pattern, labeled TAP. Tapping length increases as items proceed upward along the vertical axis from easy at the bottom to hard at the top. The number of persons scoring at each person measure is marked by a vertically printed count beneath the horizontal axis from 12 persons, who failed on all 26 patterns at measure 0, to 7 persons who succeeded on all 26 patterns at measure 100. The 2143 analyzed persons represent both genders, a wide age range, and a variety of psychological complaints.

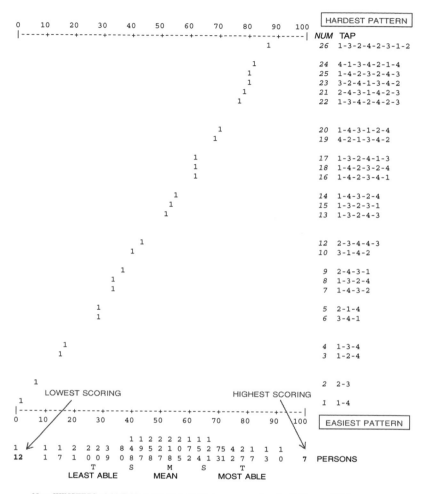

Figure 5. *KCT–R Most Probable Response Key Map*

In addition to the yardstick tapping pattern measure order, which is so clearly shown in Figure 5 on page 36 by the hierarchical trend in tapping patterns from easiest at the bottom to hardest at the top, we identify three additional features in the map that we call *line*, *stack*, and *gap* in Figure 6.

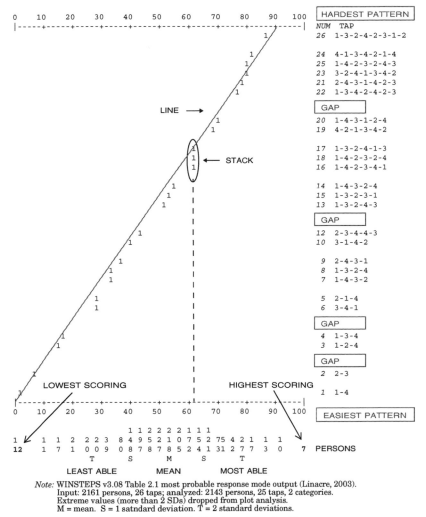

Figure 6. *KCT–R Line, Stack, Gap*

Note: WINSTEPS v3.08 Table 2.1 most probable response mode output (Linacre, 2003). Input: 2161 persons, 26 taps; analyzed: 2143 persons, 25 taps, 2 categories. Extreme values (more than 2 SDs) dropped from plot analysis. M = mean. S = 1 satndard deviation. T = 2 standard deviations.

Line. We added a straight line that reaches from lower left to upper right of Figure 6. The plot documents the extent and monotonic uniformity of the conjoint relationship between items and persons. The straighter the line, the fewer the distortions and the closer the data points to the line, the more uniform the conjoint relation between items and persons, and the clearer the definition of the metric of the yardstick that was built to define the variable.

Stack. Vertical stacks, however, mark redundancies in measure definition. There is a vertical stack of three 1s at items 16, 17 and 18, which exposes their similar difficulties at measure 62. Item stacks increase measure precision at their point of calibration, but they do not increase articulation of construct definition. Unless a measure of 62 has some particular importance, we might want to redesign one or two of these three items to see whether we can separate their calibrations along the variable and thus improve the articulation of our construct definition.

Gap. Gaps between items indicate measure regions along the line of the variable that are not defined by existing items. Gaps mark regions for which it should be possible to construct intervening items. A gap can be seen between items 12 and 13. Can we engineer one or two new items to be harder than the tapping series 2-3-4-4-3, but easier than 1-3-2-4-3? Is it the repetition of 4-4 in item 12 that makes it easier? Or is it the simple up 2-3-4 and down 4-3 progression of the pattern? Can we construct, between these two items, a new pattern of intermediate difficulty? Gaps show us where new items are implicit. The neighboring items suggest how to construct the new ones. Gaps invite us to understand our variable in more detail. Should we encounter a gap that we are unable to fill, we may have exposed a quantum step in our variable, as in Piaget's (1950) theories of stepwise intellectual development.

When item construction has addressed the line, stacks, and gaps, a second data run to collect new data for items added will show us whether we were or not successful and also where further item development is possible.[1] Item development is an important part of quality control (Stone, 2000). It sharpens and deepens our understanding of the construct we are developing.

Examination of person frequencies at each measure location shows where and how many persons are located by their measures at each point along the horizontal axis. The sample mean and one and two standard deviations in each direction away from the mean are marked M, S, and T respectively, should these points be of interest. Keep in mind, however, that these particular statistics have meaning only when the distribution of persons (or items) is approximately normal, and we are willing to think of our persons (or items) as exchangeable instances of one homogeneous population of random departures that offer no more individual information than one location (the mean) and one

---

1. For an example of this kind of construct development, see Chapter 5 of *Best Test Design* (Wright & Stone, 1979).

random, and hence, inexplicable homogeneous diversity (the standard deviation).

Figure 5 on page 36 is a map of the KCT–R variable. It shows the extent of variable construction and how well items and persons are related. To evaluate successive data analyses, we monitor whether they improve the variable features that we highlighted in Figure 6 on page 37.

Measurement order is next in importance. Table 2 provides the numerical data on which Figure 5 is based.

Table 2. *KCT–R Tap Statistics Measure Order*

ENTRY NO	RAW SCORE	COUNT	MEASURE	ERROR	INFIT		OUTFIT		SCORE CORR	TAPS
					MNSQ	ZSTD	MNSQ	ZSTD		
26	48	2143	87.1	1.0	.87	-1.1	.16	-.8	.32	1-3-2-4-2-3-1-2
24	84	2143	82.5	.8	.82	-2.0	.21	-1.0	.39	4-1-3-4-2-1-4
25	104	2143	80.6	.7	.84	-2.0	.24	-1.1	.41	1-4-2-3-2-4-3
23	106	2143	80.5	.7	.85	-1.9	.24	-1.1	.41	3-2-4-1-3-4-2
21	136	2143	78.2	.6	.96	-.6	.43	-.9	.41	2-4-3-1-4-2-3
22	158	2143	76.8	.6	.83	-2.7	.34	-1.2	.47	1-3-4-2-4-2-3
20	331	2143	69.3	.4	.93	-1.6	.56	-1.3	.53	1-4-3-1-2-4
19	367	2143	68.2	.4	.87	-3.2	.52	-1.6	.56	4-2-1-3-4-2
17	598	2143	62.1	.4	.97	-.8	.90	-.5	.58	1-3-2-4-1-3
18	601	2143	62.0	.4	.91	-2.7	.73	-1.4	.60	1-4-2-3-2-4
16	651	2143	60.9	.4	1.05	1.5	.90	-.5	.57	1-4-2-3-4-1
14	960	2143	54.7	.3	1.04	1.3	1.30	2.1	.58	1-4-3-2-4
15	1066	2143	52.7	.3	.97	-1.3	1.07	.6	.61	1-3-2-3-1
13	1108	2143	51.9	.3	1.01	.3	1.06	.5	.59	1-3-2-4-3
#12	1528	2143	43.6	.4	1.12	3.7	2.23	6.0	.50	2-3-4-4-3
10	1670	2143	40.3	.4	1.07	1.9	2.18	4.5	.50	3-1-4-2
9	1808	2143	36.3	.4	1.00	.0	1.99	2.8	.49	2-4-3-1
8	1891	2143	33.2	.5	.97	-.6	2.25	2.6	.48	1-3-2-4
# 7	1899	2143	32.9	.5	1.10	1.7	3.61	4.4	.42	1-4-3-2 ← 1-4
# 5	1997	2143	27.6	.6	1.12	1.4	5.99	4.3	.39	2-1-4 repetition
6	1997	2143	27.6	.6	.91	-1.2	1.10	.2	.45	3-4-1
4	2093	2143	17.0	1.1	.94	-.4	.48	-.5	.35	1-3-4 slow
# 3	2100	2143	15.6	1.1	1.20	1.4	9.90	3.3	.26	1-2-4 ← task
2	2128	2143	6.6	1.8	1.05	.2	1.47	.1	.18	2-3-1 adjustment
# 1	2135	2143	1.9	2.3	1.21	.7	.39	-.2	.14	1-4
MEAN	1103	2143	50.0	.7	.98	-.3	1.61	.8		
SD	795	0	24.8	.5	.11	1.7	2.11	2.2		

\# Useful MISFIT cutoffs: = 1.09   = 3.72

Note: WINSTEPS v3.08 Table 13.1 output (Linacre, 2003).
    Iinput: 2161 persons, 26 taps; analyzed: 2143 persons, 25 taps, 2 categories.
    MNSQ = mean square. ZSTD = standardized fit statistic. SCORE CORR. = correlation.

The columns from left to right list for each item the entry number, raw score, response count, item-calibration measure, calibration error, and five columns of fit statistics.

Examine the mean squares (MNSQs) in the infit column. Note the mean of 0.98 and SD (standard deviation) of 0.11 printed at the bottom of the infit column. A rough guideline for local item fit evaluation is infit values larger than one SD above the infit mean. In Table 2 on page 40, this infit guideline becomes .98 + .11 = 1.09, a value that exposes possible infit misfit in items 1, 3, 5, 7, and 12. The misfit values for the four items (1, 3, 5, 7) early in the series are probably due to attention lapses among persons who are slow to adapt to the task. The perhaps unexpected tapping repetition of 4-4 in item 12 may be the cause of the misfit of this item.

Outfit values are also useful. Recall the difference between infit and outfit. The outfit guideline in Table 2 on page 40 is 1.61 + 2.11 = 3.72. Only items 3 and 5 show outfits in excess of 3.72. Item 3 is the first 3-tap item encountered and item 5 is the first item that begins the tapping series by moving down to the left instead of up to the right. Perhaps this surprise in the tapping sequence has disrupted some persons. We can identify these persons when we are ready to study them.

Finally, the SCORE CORR (correlation) column gives the correlations between person-measure and person-response for each item. Because all correlations are positive, there is no polarity problem in these data. As usual the correlations are highest in the middle, where there is the most variance, and lowest at the top and bottom of the tapping series, where there is the least variance.

Figure 7 plots the principal components (standardized residuals) analysis of item response residuals similarities against their difficulty calibrations.

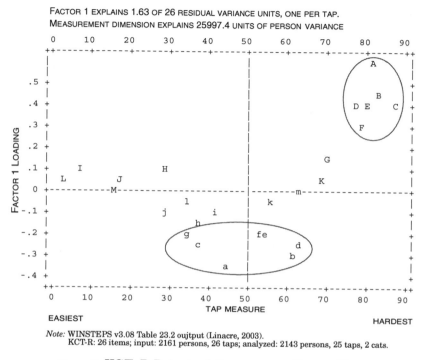

FACTOR 1 EXPLAINS 1.63 OF 26 RESIDUAL VARIANCE UNITS, ONE PER TAP.
MEASUREMENT DIMENSION EXPLAINS 25997.4 UNITS OF PERSON VARIANCE

Note: WINSTEPS v3.08 Table 23.2 oujtput (Linacre, 2003).
      KCT-R: 26 items; input: 2161 persons, 26 taps; analyzed: 2143 persons, 25 taps, 2 cats.

Figure 7. *KCT–R Principal Components Factor Plot*

This plot shows item residual factor loadings on the vertical axis plotted against item difficulty calibrations on the horizontal axis. Substantial factor loading deviations from 0 invite investigation, especially when groups of items cluster together. In this plot we see items labeled A, B, C, D, E, and F clustered in the upper right-hand portion of the plot.

The table of factor loadings, Table 3, lists the items to which these labels refer.

Table 3. *Factor 1 from KCT–R Principal Components Analysis*

FACTOR	LOADING	MEASURE	INFIT MNSQ	OUTFIT MNSQ	ENTRY NO.	TAP
1	.59	80.7	.84	.24	A–25	1-4-2-3-2-4-3
1	.44	82.6	.82	.22	B–24	4-1-3-4-2-1-4
1	.42	87.1	.86	.16	C–26	1-3-2-4-2-3-1-2
1	.38	77.0	.82	.33	D–22	1-3-4-2-4-2-3
1	.38	80.6	.85	.24	E–23	3-2-4-1-3-4-2
1	.29	78.3	.95	.43	F–21	2-4-3-1-4-2-3
1	.14	69.5	.92	.55	G–20	1-4-3-1-2-4
1	.11	28.5	.92	1.05	H– 6	3-4-1
1	−.37	44.3	1.08	2.01	a–12	2-3-4-4-3
1	−.30	61.2	1.03	2.01	b–16	1-4-2-3-4-1
1	−.23	62.4	.97	88	d–17	1-3-2-4-1-3
1	−.21	53.2	.95	1.04	e–15	1-3-2-3-1
1	−.18	52.4	1.01	1.05	f–13	1-3-2-4-3
1	−.18	33.8	1.09	2.91	g– 7	1-4-3-2
1	−.14	37.2	1.00	1.84	h– 9	2-4-3-1
1	−.11	41.1	1.05	2.01	i–10	3-1-4-2
1	−.10	28.5	1.14	5.16	j– 5	2-1-4

bold = HARDEST PATTERNS

Note: WINSTEPS v3.08 Table 23.2, Factor 1 oujtput (Linacre, 2003).
KCT-R: 26 items; input: 2161 persons, 26 taps; analyzed: 2143 persons, 25 taps, 2 cats.
Standardized Residual Correlations for TAPS (Sorted By Loading)
Factor 1 explains 1.63 of 26 residual variance units, one per TAP.
Measurement dimension explains 25997.4 units of PERSON variance.

We can use Table 3 to study the loadings and measures for these items together with their infit and outfit values and their tapping patterns. The items labeled A to F in the plot refer to items 21 to 26, the six most difficult tapping patterns. A second group of items (a, b, d, e, and f), with negative loadings clustered at the bottom of the plot, are items

12, 14, 15, 16, and 17 located in the middle of the tapping pattern calibrations.

What do these tapping pattern clusters suggest? Part of the answer is found above the top of the plot in the Figure 7 note that reads, "Factor 1 explains 1.63 of 26 residual variance units, one per tap." The yardstick dimension explains 25997.4 equivalent units of person variance. The factor sensitivity ratio here is 1.63 / 25997.4 = .00006, which is minuscule. This answers whether the identified clusters threaten the stability of the KCT–R yardstick: obviously not.

Principal components analysis sensitizes us to possible areas of concern. We see, in this case, that there is no merit to our concern. Erratic individual performance was found among a few persons, which is the source of item outfit. Clinicians can use the misfit detection to identify attention span lapses, which is the diagnostic purpose of the KCT–R. We concentrate first on the development of the measuring instrument, the yardstick. Consequently, at this time we need only determine the extent to which erratic persons interfere with yardstick development. The results in Figure 7 show that they do not.

Erratic persons are identified in Figure 4. This output table shows a Guttman pattern of most unexpected responses by item and persons. Responses are marked as unexpected when the odds against their occurrence are at least 4 to 1. here we see in detail exactly who caused items 1, 3, 5, 7, and 12 to be identified as misfitting.

Table 4. *KCT–R Most Unexpected Response Guttman Diagram*

TAP	MEASURE	PERSON

| | | HIGH PERSON | most erratic persons | LOW PERSON |

```
                              HIGH          most erratic persons        LOW
                              PERSON                                    PERSON
                              11111 211111  111  1111111    111  1111122  1 1
                              58353207553332862448552994443222121988411038716266
                              60141496440048449834108108325610598087350143985620
                              03714774639583299105485421515883451914584746856583
                       HIGH------------------------------------------------
  1  1-4         1.9  J   ...............................................00
  2  2-3         6.6  H   .................................................0..00000.
  3  1-2-4      15.6  A   .....0................................0..0..0..0.....
  4  1-3-4      17.0  i   .................................................0......
  5  2-1-4      27.6  B   .....0...........00.0..........0.0.0.0...........
  6  3-4-1      27.6  K   ........................0.0....00...0.....0
  7  1-4-3-2    32.9  C   ...0...........0.0..0... 0000.................0....
  8  1-3-2-4    33.2  D   ...........0....0....00...0..............0.....
  9  2-4-3-1    36.3  G   ...........0 0...00. .................0.....
 10  3-1-4-2    40.3  F   ......0000.0...........0.................0....
 12  2-3-4-4-3  43.6  E   ..0.0.....0... .0.........................
 13  1-3-2-4-3  51.9  M   ..............0....00.0..........11....
 15  1-3-2-3-1  52.7  L   ......0......0 ..........0..................
 14  1-4-3-2-4  54.7  I   00....... ..........00.....................
 16  1-4-2-3-4-1 60.9 l   ...........0....0..0............11...1...
 18  1-4-2-3-2-4 62.0 g   ........0..0.....................1.1.....
 17  1-3-2-4-1-3 62.1 k   .......0.........................1.........
 20  1-4-3-1-2-4 69.3 h   ...........................1............
 22  1-3-4-2-4-2-3 76.8 b .0...........................1..............
 21  2-4-3-1-4-2-3 78.2 j .....................1..111.1....1...........
 23  3-2-4-1-3-4-2 80.5 d .................1111....1..............
 25  1-4-2-3-2-4-3 80.6 c ................11...11..11............
 24  4-1-3-4-2-1-4 82.5 a ...........11.........11................
 26  1-3-2-4-2-3-1 87.1 e ......1...1.1............1...1...............
                       ---------------------------------------------------LOW-
                              11111221111132111441111114432111211111223811266
                              58353407553348862838552994325622198988411043786620
                              60141796440083449104108108515810551087350146955583
                              03714 746395  299  5485421    834  9145847  8 6
```

Note: Only persons #1842 and #1481 show three lapses.
WINSTEPS v3.08 Table 10.5 most erratic persons output (Linacre, 2003).
Input: 2161 persons, 26 taps; analyzed: 2143 persons, 25 taps, 2 categories.

Unexpected failures on these easy items appear as 0 at the top of the Table 4; less frequent unexpected successes appear as 1 at the bottom. Cells that contain a " . " did not contain unexpected responses. This output shows us which persons might be sufficiently erratic to be deleted before completing our yardstick definition. It also shows us exactly who is manifesting attention lapses that may be diagnostically important for them as individuals. In fact, only 2 persons show as many as 3 unexpected lapses.

When we desire more person information, the unexpected lapses output in Figure 8 on page 47 gives us the detail we need. We selected persons 416, 268, and 514, males aged 45, 30, and 52 respectively for illustration. They are the 3 most misfitting responses patterns among these 2161 persons. Their response sequences show the occurrence of their unexpected misfits. We can see exactly which tapping series they missed. Because successful guessing is impossible, misfit on the KCT–R yardstick indicates lapses of attention. The 0s mark these lapses. The 1s indicate successes that become unexpected because prior lapses have lowered the person's overall measure.

Person 416 has an overall measure of 36.0. This measure is lowered by the string of four lapses on tapping patterns 2, 3, 4, 5. Perhaps he misunderstood the task. The consequences of this string of four lapses is to make success on advanced items 21, 25, and 24 sufficiently unexpected to be marked (1). Parentheses indicate that if this person's measure is 36, the odds against these particular unexpected successes exceed 4 to 1. Notice that the presence of these aberrations is also indicated by an infit statistic of 4.3 and an outfit statistic of 9.9.

```
PER    NAME    MEASURE    INFIT   (MNSQ)    OUTFIT    S.E.
416    m45     36.0       4.3       G        9.9      4.6    NUM    TAP
0    10    20    30    40    50    60    70    80    90    100
|-----+-----+-----+-----+-----+-----+-----+-----+-----+-----|
                          0                (1)                      14    1-4-3-2-4
    UNEXPECTED            0                (1)                      15    1-3-2-3-1
      LAPSES             0                (1)                      13    1-3-2-4-3
                          0       .1.                              12    2-3-4-4-3
            (0)          1                                          5    2-1-4
      (0)                1                                          4    1-3-4
      (0)                1                                          3    1-2-4
0)                       1                                          2    2-3
|-----+-----+-----+-----+-----+-----+-----+-----+-----+-----|
0    10    20    30    40    50    60    70    80    90    100


PER    NAME    MEASURE    INFIT   (MNSQ)    OUTFIT    S.E.
268    m30     66.2       3.2       C        9.9      4.6    NUM    TAP
0    10    20    30    40    50    60    70    80    90    100
|-----+-----+-----+-----+-----+-----+-----+-----+-----+-----|
                                        0            (1)           26    1-3-2-4-2-3-1-2
    UNEXPECTED                          0         (1)              25    1-4-2-3-2-4-3
      LAPSES                           0         (1)              23    3-2-4-1-3-4-2
                                        0    .1.                   20    1-4-3-1-2-4
                              .0.       1                          17    1-3-2-4-1-3
                              .0.       1                          16    1-4-2-3-4-1
                        (0)             1                          15    1-3-2-3-1
            (0)                         1                           8    1-3-2-4
|-----+-----+-----+-----+-----+-----+-----+-----+-----+-----|
0    10    20    30    40    50    60    70    80    90    100


PER    NAME    MEASURE    INFIT   (MNSQ)    OUTFIT    S.E.
514    m52     75.6       3.1       S        3.7      4.4    NUM    TAP
0    10    20    30    40    50    60    70    80    90    100
|-----+-----+-----+-----+-----+-----+-----+-----+-----+-----|
                                        0            (1)           26    1-3-2-4-2-3-1-2
    UNEXPECTED                          0    .1.                   24    4-1-3-4-2-1-4
      LAPSES                           0    .1.                   25    1-4-2-3-2-4-3
                                        0     .1.                  21    2-4-3-1-4-2-3
                                        0    .1.                   22    1-3-4-2-4-2-3
                              .0.       1                          20    1-4-3-1-2-4
                        (0)             1                          17    1-3-2-4-1-3
                        (0)             1                          16    1-4-2-3-4-1
                  (0)                   1                          13    1-3-2-4-3
|-----+-----+-----+-----+-----+-----+-----+-----+-----+-----|
0    10    20    30    40    50    60    70    80    90    100
```

Note: WINSTEPS v3.08 Table 7.3 output (Linacre, 2003).
KCT-R: 26 items; input: 2161 persons, 26 taps; analyzed: 2143 persons, 25 taps, 2 cats.
0s = lapses. 1s = successes that become unexpected because of prior lapses.

Figure 8. *KCT–R Unexpected Lapses*

When we identify the specific inattention of these persons and others like them, we might not include their data in the yardstick construction calibration. We would certainly include them later, after the definition of our KCT-R yardstick is established, in order to do them the justice of a potentially useful personal diagnosis of attention span vulnerability.

Reliability

Test reliability is commonly reported as a correlation index R. Although R is widely used, its characteristics are widely misunderstood. The increments between successive values are not equal, so they cannot produce exchangeable interpretations. The increment between .8 and .85 is not nearly as large as the increment from .9 to .95, and both are substantially greater than the increment from .5 to .55. Nonlinearity makes these coefficients intractable for use in arithmetic operations. To become linear, correlation coefficients must be converted to Fisher z scores.

This problem can be fixed, however, if we use the square root of reliability R divided by (1 - R) to define a separation S:

$$S = [R/(1 - R)]^{\frac{1}{2}} = SDA_{sample}/SD_{error} \qquad (19)$$

where SDA_{sample} is an error-corrected sample standard deviation as in

$$SDA^2{}_{sample} = SD^2{}_{sample} - (RMSE_{error})^{\frac{1}{2}} \qquad (20)$$

and $RMSE_{error}$ is the root mean square error of person measures.

Separation (S) expresses reliability as a ratio of the error-corrected sample standard deviation of the persons to the root mean square test error of person measures, a sample property in the numerator compared to a test property in the denominator. Separation values and their relationship to reliabilities (R) can be seen in Table 5.

Table 5. *Separation (S) Reliability (R) Relationships*

Separation	SDA	SEM	R
1	1	1.00	.50
2	1	.50	.80
3	1	.33	.90
4	1	.25	.94

When S = 1, persons are no more spread than the uncertainty in their test measures. The two distributions are the same. There is no way to distinguish one person from another. But if S = 3, then persons are 3 times more spread than their root mean square test error. We see in the pictures of the two distributions that as separation increases, it becomes easier to distinguish among persons. A separation of 3 is equivalent to a reliability of .90. Now, at last, we have a way to explain the meaning of a reliability of .90.

Table 6 summarizes the statistics for measured persons and taps.

Table 6. *Summary of Measured Persons and Taps*

Summary of 2143 nonextreme measured persons

	RAW SCORE	COUNT	MEASURE	MODE ERROR	INFIT		OUTFIT	
					MNSQ	ZSTD	MNSQ	ZSTD
MEAN	12.9	25.0	52.56	4.54	.98	-.3	.90	-.1
SD	3.8	0	13.12	.29	.56	1.3	1.72	.2
MAX	24.0	25.0	91.69	7.41	4.52	4.8	9.90	2.7
MIN	1.0	25.0	1.72	4.31	.28	-2.3	.07	-.5

Note: Real RMSE = 5.00, Adj.SD = 12.13, S = 2.43, Person Reliability = .85
S.E. of Person Mean = .28
WIth 18 Extreme = 2161, Persons, Mean = 52.40, SD = 14.01
Real RMSE = 5.09, Adj.SD = 13.05, Separation = 2.56, Person Relability = .87

Summary of 25 measured taps

	RAW SCORE	COUNT	MEASURE	MODE ERROR	INFIT		OUTFIT	
					MNSQ	ZSTD	MNSQ	ZSTD
MEAN	1102.6	2143.0	50.00	.69	.98	-.3	1.61	.8
SD	794.6	0	24.76	.46	.11	1.7	2.11	2.2
MAX	2135.0	2143.0	87.06	2.31	1.21	3.7	9.90	6.0
MIN	48.0	2143.0	1.92	.34	.82	-3.2	.16	-1.6

Note: WINSTEPS v3.08 Table 3.1 output (Linacre, 2003). Deleted: 1 Tap
Real RMSE = .87, Adj.SD 24.74, Separation = 28.53, Reliability = 1.00
S.E. of Tap Mean = 5.05

The lower section of Table 6 reports an item separation S of 28.53 (a Cronbach Alpha or KR-20 reliability equivalent of 1.00). This documents that the operational definition of the tapping yardstick is extremely clear and articulate. The top section of Table 6 reports a person separation S of

2.43 (reliability of .85), which is lower. If we decide to compare these separation statistics, we must adjust their ratio
for their differences in replication,

$$(\text{sep}_p/\text{sep}_i) \;/\; (i/p)^{\frac{1}{2}}. \tag{21}$$

For our data, where $i = 26$ and $p = 2143$, the
adjusted ratio becomes

$$(2.43/28.53) \;/\; [(26/2143)]^{\frac{1}{2}} = 0.77. \tag{22}$$

Were these two separation statistics equal, the ratio
would be $1:1$. In this example the adjusted person separation is less than the adjusted item separation by a factor of
0.77. The KCT–R yardstick is more replicable than the performance of persons, perhaps because of the attention lapses
that were previously identified.

Chapter 8   *Fear Survey Schedule*

This chapter illustrates the power of applying measurement principles to improve the measure by examining specific rating scale functioning and item characteristics. We want to know how well individual respondents express themselves with the rating scale on the items, and we want to know how well each item is working.

The *Fear Survey Schedule* (Wolpe & Lang, 1964) lists 108 possibly fearful situations. Item examples include

     1. "Noise of vacuum cleaners,"
     16. "Failure,"
     19. "Looking down from high buildings,"
     52. "Being in an elevator,"
     66. "Cemeteries," and
     96. "Hurting the feelings of others."

Respondents rate the extent of their fear using

0 = Not at all,
1 = A little,
2 = A fair amount,
3 = Much, or
4 = Very much.

The 223 persons were clients of an outpatient mental health facility. Table 7 on page 54 provides the *Fear Survey Schedule* data.

Table 7. *Fear Survey Data: 5-Category Rating Scale Results*

Summary of 223 measured persons

	RAW SCORE	count	measure	MODE ERROR	INFIT		OUTFIT	
					MNSQ	ZSTD	MNSQ	ZSTD
MEAN	97.4	107.8	38.95	1.14	1.08	0	1.07	0
SD	58.5	.6	6.83	.43	.52	2.9	.55	2.7

Real RMSE 1.38, Adj.SD 6.69, Separation 4.83, Person Reliability .96

Summary of 108 measured fears

	RAW SCORE	COUNT	MEASURE	MODE ERROR	INFIT		OUTFIT	
					MNSQ	ZSTD	MNSQ	ZSTD
MEAN	201.0	222.6	50.00	.77	1.10	.5	1.07	.4
SD	99.4	.7	5.99	.21	.25	2.1	.29	2.2

Real RMSE .87, Adj.SD 5.92, Separation 6.77, Fear Reliability .98

Summary of measured steps (5 category labels)

CAT LABEL	SCORE	OBSV. COUNT	%	OBSV AVERAGE	SAMPLE EXPECT	INFIT MNSQ	OUTFIT MNSQ	STEP CALIBRATION	CATEGORY MEASURE
0	0	11881	49	−16.3	−15.7	.94	.3	none	(−19.67)
1	1	6288	26	−8.7	−9.9	.99	.75	−6.87	−7.15
2	2	3199	13	−4.7	−5.3	.94	.88	−1.39	.19
3	3	1685	7	−1.6	−1.3	1.04	1.12	2.55	7.31
4	4	993	4	−.3	2.5	1.42	2.07	5.71	(19.08)
MISSING		38	0	−6.5					

0 = Not at all,
1 = A little,
2 = A fair amount,
3 = Much
4 = Very much.

Note: WINSTEPS v3.08 Table 3.1 summaries of persons and items output (Linacre, 2003).
Input: 223 persons, 108 fears.
Analyzed: 223 persons, 108 fears, 5 categories.

How Many Rating Scale Categories?

First, we examine how the respondents were able to use the 5-category rating scale to express useful information. Table 7 reports a person separation S index of 4.83 (R = .96) and an item separation S index of 6.77 (R = .98). These statistics document a high level of instrument consistency, that is, the presence of a stable yardstick for measuring respondents' fear. However, there is more to the story. How well are these five rating scale categories working? Can this rating structure be improved?

The information about how well these categories are working is located at the bottom of Table 7. Here, in the summary of measured steps left-most frame, each category label (0, 1, 2, 3, 4) is accompanied by its observed count and percentage. This documents the popularity of each category. The next frame to the right gives the observed and expected average measures in each category. These are the average person measures that accompany each occasion on which a category is chosen. Further to the right are the infit and outfit statistics for each category. These fit values show that only category label "4 = Very much" manifests a discrepant infit of 1.42 and outfit of 2.07.

Because the fit statistics for category "4 = Very much" show that it has drawn many unexpected choices, we combine categories "3 = Much" and "4 = Very much" to see whether that would improve the functioning of this rating scale (see Table 8 on page 56).

Table 8. *Fear Survey Data: Combining Rating Categories 3 and 4*

Summary of 223 measured persons

	RAW SCORE	COUNT	MEASURE	MODE ERROR	INFIT		OUTFIT	
					MNSQ	ZSTD	MNSQ	ZSTD
MEAN	92.9	107.8	41.51	1.22	1.05	-.1	1.05	0
SD	53.5	.6	7.48	.42	.41	2.6	.47	2.4

Real RMSE 1.43, Adj.SD 7.34, Separation 5.12, Person Reliability .96

Summary of 108 measured fears

	RAW SCORE	COUNT	MEASURE	MODE ERROR	INFIT		OUTFIT	
					MNSQ	ZSTD	MNSQ	ZSTD
MEAN	191.9	222.6	50.00	.83	1.05	.3	1.05	.4
SD	93.1	.7	6.57	.19	.19	1.9	.25	1.9

Real RMSE .91, Adj.SD 6.50, Separation 7.18, Fear Reliability .98

Summary of measured steps (combining category labels 3 and 4)

CAT LABEL	SCORE	OBSV. COUNT	%	OBSV AVERAGE	SAMPLE EXPECT	INFIT MNSQ	OUTFIT MNSQ	STEP CALIBRATION	CATEGORY MEASURE
0	0	11881	49	-14.3	-13.8	.94	.3	none	(-19.67)
1	1	6288	26	-5.9	-7.1	.99	.75	-6.87	-7.15
2	2	3199	13	-1.3	-1.6	.94	.88	-1.39	.19
3	3	2678	11	2.4	3.7	1.19	1.44	2.71	(16.59)

0 = Not at all,
1 = A little,
2 = A fair amount,
3 = Much and 4 = Very much.

Note: WINSTEPS v3.08 Table 3.1 summaries of persons and items output (Linacre, 2003).
Input: 223 persons, 108 fears.
Analyzed: 223 persons, 108 fears, 4 catetories.

As a result of combining categories to in effect create a virtual 4-category rating scale, the above table shows an improved person separation index of 5.12 (R = .96) and an improved item separation index of 7.18 (R = .98).

These values are greater than the values for the 5-category rating scale, which were 4.83 and 6.77. Combining categories 3 and 4 produced a gain in response consistency and thus increased information flow.

For this combination, the infit mean square has reduced to 1.19 and the outfit mean square to 1.44. Combining categories 3 and 4, which omits reckoning a consistent difference in meaning between "Much" and "Very Much," has improved the signal-to-noise ratio of our fear yardstick.

For these 223 persons, the increase in clarity that results from combining "3 = Much" and "4 = Very much" is visually apparent when we compare their probabilities of response step measures at intersections curves of Figures 9 and 10.

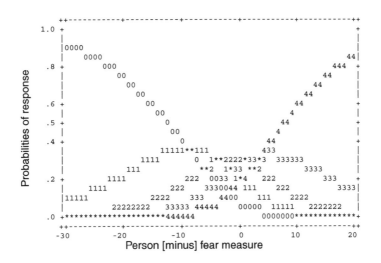

Figure 9. *Fear Survey Data 5-Category Curves*

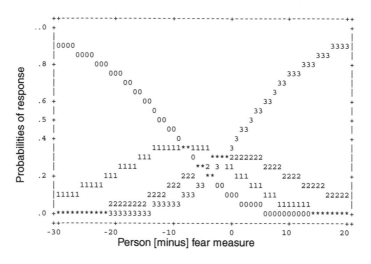

Figure 10. *Fear Survey Data 4-Category Curves*

This reflects what we see in Table 9 on page 59: when the number of categories is further reduced to only three (by combining "A fair amount," "Much," and "Very Much"), person separation increases even more to 5.53 (R = .97), and item separation to 7.67 (R = .98).

Table 9. *Fear Survey Data: Combining Rating Categories 2, 3 and 4*

Summary of 223 measured persons

	RAW SCORE	COUNT	MEASURE	MODE ERROR	INFIT		OUTFIT	
					MNSQ	ZSTD	MNSQ	ZSTD
MEAN	80.9	107.8	44.42	1.48	1.03	-.1	1.05	-.1
SD	41.8	.6	9.25	.41	.30	2.2	.43	2.0

Real RMSE 1.65, Adj.SD 9.11, Separation 5.53, Person Reliability .97

Summary of 108 measured fears

	RAW SCORE	COUNT	MEASURE	MODE ERROR	INFIT		OUTFIT	
					MNSQ	ZSTD	MNSQ	ZSTD
MEAN	167.1	222.6	50.00	1.01	1.01	0	1.05	.4
SD	77.5	.7	8.16	.18	.14	1.7	.22	1.6

Real RMSE 1.65, Adj.SD 9.11, Separation 5.53, Fear Reliability .97

Summary of measured steps (combining category labels 2, 3, and 4)

CAT LABEL	SCORE	OBSV. COUNT	%	OBSV AVERAGE	SAMPLE EXPECT	INFIT MNSQ	OUTFIT MNSQ	STEP CALIBRATION	CATEGORY MEASURE
0	0	11881	49	-12.9	-12.7	.97	1.03	none	(-19.67)
1	1	6288	26	-2.1	-3.2	.89	.86	-1.97	0
2	2	5877	24	5.5	6.1	1.09	1.21	1.97	(14.82)

0 = Not at all,
1 = A little,
2 = A fair amount, 3 = Much, and 4 = Very much.

Note: WINSTEPS v3.08 Table 3.1 summaries of persons and items output (Linacre, 2003).
Input: 223 persons, 108 fears. Analyzed: 223 persons, 108 fears, 3 categoreis.

For this combination of all three top categories (3, 4, and 5), the infit mean square reduced to 1.09, and the outfit mean square to 1.21. At this point, we wonder what will happen if we make the scale a dichotomy and only note whether the respondent has replied "Not at all" or even "A little"?

Table 10 presents dichotomous results. This reduction in categories does not increase person separation. Instead, person separation dropped to 5.12 (R = .96) and item separation to 7.25 (R = .98). With a dichotomy, the infit mean square is .99 and the outfit mean square 1.03.

Table 10. *Fear Survey Data: Dichotomy*

Summary of 223 measured persons

	RAW SCORE	COUNT	MEASURE	MODE ERROR	INFIT MNSQ	INFIT ZSTD	OUTFIT MNSQ	OUTFIT ZSTD
MEAN	80.9	107.8	44.42	1.48	1.03	-.1	1.05	-.1
SD	41.8	.6	9.25	.41	.30	2.2	.43	2.0

Real RMSE 1.65, Adj.SD 9.11, Separation 5.53, Person Reliability .97

Summary of 108 measured fears

	RAW SCORE	COUNT	MEASURE	MODE ERROR	INFIT MNSQ	INFIT ZSTD	OUTFIT MNSQ	OUTFIT ZSTD
MEAN	167.1	222.6	50.00	1.01	1.01	0	1.05	.4
SD	77.5	.7	8.16	.18	.14	1.7	.22	1.6

Real RMSE 1.06, Adj.SD 8.10, Separation 7.67, Fear Reliability .98

Summary of measured steps (combining category labels 1, 2, 3, and 4)

CAT LABEL	SCORE	OBSV. COUNT	%	OBSV AVERAGE	SAMPLE EXPECT	INFIT MNSQ	OUTFIT MNSQ	COHERHENCE M →C	COHERHENCE C →M
0	0	11881	50	-10.6	-10.6	.97	1.20	77%	77%
1	1	11951	50	-2.1	-3.2	.89	.86	77%	77%
MISSING		38	0	12.7					

0 = Not at all,
1 = A little, 2 = A fair amount, 3 = Much, and 4 = Very much.
M →C = Does the measure implies a category?
C →M = Does the category implies a measure?

Note: WINSTEPS v3.08 Table 3.1 summaries of persons and items output (Linacre, 2003).
Input: 223 persons, 108 fears. Analyzed: 223 persons, 108 fears, 2 categoreis.

We can visually compare the results in Figures 11 and 12.

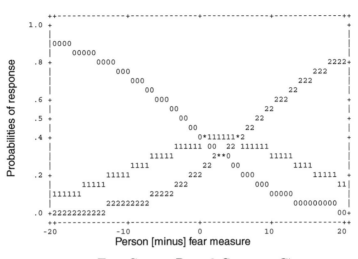

Figure 11. *Fear Survey Data 3-Category Curves*

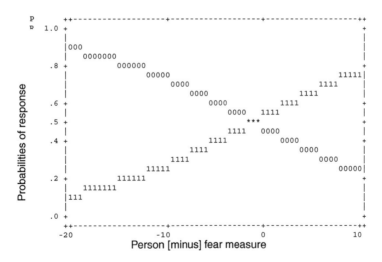

Figure 12. *Fear Survey Data Dichotomous Curves*

Figures 9 through 12 and Tables 7 through 10 provide results for different combinations of rating scale categories. This succession of category reductions shows how these persons use the fear rating scale. A trichotomy that includes "Not at all," "A little," and "A fair amount or more" produced the best signal-to-noise ratio. Table 11 provides the values for the original 5-category rating scale, a 4-category rating scale, its reductions to the 3-category, and to a dichotomy.

Table 11. *Rating Scale Reduction*

RATING SCALE	SEPARATION		RELIABILITY		INFIT MNSQ	OUTFIT MNSQ
	PERSON	ITEM	PERSON	ITEM		
5-Category	4.83	6.77	.96	.98	1.42	2.07
4-Category	5.12	7.18	.96	.98	1.19	1.44
3-Category	**5.53**	**7.67**	**.97**	.98	1.09	1.21
Dichotomy	5.12	7.75	.96	.98	.99	1.03

Combining the responses of the 5-category fear scale to a dichotomy increases the separation statistics and reduces misfit! Only the 3-category version (results in bold) exceeds the dichotomous separation and reliability statistics. Of what value then is the original 5-category rating scale? Statistically, even a dichotomy is better than the original 5-category scale. Does this mean that we need to change the printed form of the fear survey? Not necessarily. We can continue to use the original form for recording assessments, if that pleases respondents. But we will make our measures from the strongest model, a trichotomy, or use the simplest model, a dichotomy, because it is the most efficient for constructing and reporting measures.

Rating scale expansion is often recommended to "increase the variance." The assumption behind this advocacy is that increasing the number of categories collects more data and hence automatically improves measurement. However, we need to investigate and confirm our presumptions for improving measurement. If more categories are provided in the survey form but not used as intended by respondents, then "more" categories are not operating to collect more information. Unless the unproductive additional categories are making the response task easier for respondents, it is more useful to reduce the number of response categories offered to the number that is actually needed for measuring. In the case of the fear survey, two or three categories are all that are needed to construct the best measures.

There are many reasons why this can occur. Respondents may not need the gradations of the rating scale, but actually experience the item as a dichotomy, a "yes/no," and respond accordingly. In his analysis of Wolpe and Lang's (1964) fear scale, Stone (1998) showed that a dichotomous model was as satisfactory as the full 5-point rating scale. Stone's (1998) study of the Beck Depression Inventory also showed that a dichotomy was a better model for measurement than Beck's 4-point rating scale.

Item Dimensions

Next, we investigate how specific items are working with the measure. As shown in Figure 13, the fear data reveal that item principal components analysis explains only 5.9 of 108 standardized residual variance units, and the fear measurement dimension explains 115.3 units. This ratio 5.9:115.3 indicates a negligable 5% disturbance in the fear yardstick. Even so, the substantive split is interesting.

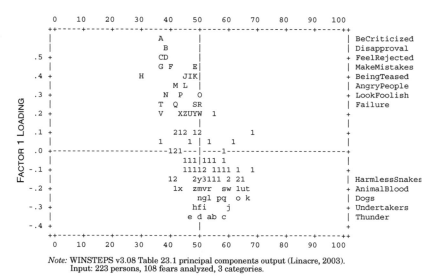

Figure 13. *Fear Response Data Principal Components*

Table 12 on page 65 shows that principal components analysis factor 1 positive loadings identify internal social fears such as invoking shame and the negative loadings identify external things suggesting danger.

Table 12. *Fear Survey Principal Components Factor 1 Loadings*

FACTOR	LOADING	MEASURE	OUTFIT MNSQ	INFIT MNSQ	CD	NUM	FEAR
1	.59	36.7	.94	1.00	A	49	BeingCriticized
1	.53	37.6	.83	1.05	B	64	Disapproval
1	.50	36.2	.84	.88	C	61	FeelingRejected
1	.49	38.8	.85	.91	D	75	MakingMistakes
1	.45	48.1	.99	1.02	E	12	BeingTeased
1	.44	40.5	.87	1.58	F	54	AngryPeople
1	.43	37.4	.91	1.26	G	76	LookingFoolish
1	.41	30.2	1.08	1.31	H	16	Failure
1	.39	46.1	1.05	1.14	I	26	FeelingAngry
1	.38	46.7	1.01	1.19	J	27	Authorities
1	.38	47.6	1.24	1.28	K	4	LoudVoices
1	.36	44.2	.92	.90	L	67	BeingIgnored
1	-.36	53.7	1.03	1.08	a	14	Thunder
1	-.35	55.6	1.05	.99	b	98	Undertakers
1	-.34	57.8	1.06	.91	c	87	Dogs
1	-.33	49.9	.82	.77	d	57	AnimalBlood
1	-.33	47.3	1.19	1.17	e	65	HarmlessSnakes
1	-.31	47.8	.99	.94	f	56	HumanBlood
1	-.31	50.0	.99	.98	g	80	HarmlessSpiders
1	-.29	48.1	.90	.90	h	44	CrawlingInsects
1	-.29	51.7	.96	1.06	i	72	Lightening
1	-.29	60.8	.92	.81	j	43	BusJourneys
1	-.27	66.1	1.00	.66	k	42	Dirt
1	-.27	54.0	.99	.89	l	66	Cemeteries

Note: WINSTEPS v3.08 Table 23.2 standardized residuals output (Linacre, 2003).
Input: 223 persons, 108 fears analyzed, 3 categories.
CD = principal components plot code.
NUM = entry number.

We can use this information diagnostically to differentiate between internal and external fears, thus increasing the clinical utility of the fear scale.

We could also edit the fear scale by reducing redundant items. Table 13 reveals that item 56, "human blood" and 57, "animal blood" are very similar in impact. So are item 81, "making decisions," and 107, "being in charge." Likewise for the remaining pairs. If we want to shorten the instrument, this table of residual item correlations provides guidelines for how to do so with the least loss of information. We can use the best fitting of each pair instead of both.

Table 13. *Fear Survey Largest Standardized Residual Correlations*

RESIDUAL CORRELATION	NUM	FEAR	NUM	FEAR
.63	56	HumanBlood	57	AnimalBlood
.60	81	MakingDecisions	107	BeingInCharge
.60	18	HighPlaces	19	LookngDwnFrmHgh
.54	14	Thunder	72	Lightening
.50	11	Autos	31	CarJourneys
.45	50	StrangeShapes	90	Medicine
.45	78	Fainting	79	Nausea
.44	35	ToSeeBullying	45	SeeingAFight
.44	28	FlyingInsects	44	CrawlingInscts
.44	31	CarJourneys	86	OpenSpaces

Note: WINSTEPS v3.08 Table 23.2 standardized residuals output (Linacre, 2003). Input: 223 persons, 108 fears analyzed, 3 categories.

Table 14 on page 67 shows items in descending order of infit. The better fitting items, infit mean of 1.01 with a SD of .14, provide a core stability for the fear scale. Infit values greater than $1.01 + .14 = 1.15$ suggest items that may be too vague and unclear for reproducible measurement. The best fitting items at the bottom of Table 14 are more commonplace fears. We might want to build a shortened measuring scale based on just these most useful items.

Table 14. *Fear Survey Statistics Infit Order*

NUM	RAW SCORE	COUNT	MEASURE	ERR.	INFIT MNSQ	OUTFIT MNSQ	CD	SCORE CORR	FEARS
93	113	221	54.3	1.0	1.45	1.55	A	.40	Homosexuality
92	159	223	49.9	.9	1.40	1.49	B	.38	GodsPunishment
104	156	223	50.2	.9	1.37	1.44	C	.44	Marriage
13	257	223	41.3	.9	1.31	1.35	D	.40	Dentists
18	180	222	47.9	.9	1.29	1.36	E	.45	HighPlaces
5	265	222	40.5	.9	1.28	1.22	F	.43	DeadPeople
19	252	223	41.7	.9	1.26	1.36	G	.38	LookingDwnFrmHigh
4	184	223	47.6	.9	1.24	1.28	H	.44	LoudVoices
6	306	223	36.9	.9	1.23	1.65	I	.35	SpeakingInPublic
59	169	222	48.8	.9	1.19	1.25	J	.48	Enclosures
65	187	223	47.3	.9	1.19	1.17	K	.49	HarmlessSnakes
3	160	223	49.8	.9	1.18	1.76	L	.41	BeingAlone
84	237	223	43.0	.9	1.17	1.17	M	.44	TakingTests
38	130	223	52.8	1.0	1.16	1.31	N	.48	DeepWater
34	53	222	63.6	1.4	1.16	1.10	O	.45	Cats
10	252	223	41.7	.9	1.16	1.22	P	.45	Falling
101	67	219	60.4	1.2	1.16	1.19	Q	.46	Masturbation
24	230	223	43.6	.9	1.15	1.27	R	.47	Bats
70	99	221	56.1	1.1	1.14	1.23	S	.50	NudeMen
83	198	223	46.4	.9	1.14	1.16	T	.50	MentalIllness
105	136	223	52.1	1.0	1.12	1.12	U	.52	Insecticides
2	254	222	41.4	.9	1.12	1.37	V	.38	OpenWounds
22	216	223	44.8	.9	1.12	1.24	W	.41	Injections
106	167	222	49.1	.9	1.11	1.10	X	.50	Vomiting
53	243	222	42.4	.9	1.11	1.16	Y	.46	SeeingSurgery
47	271	222	40.0	.9	1.10	1.27	Z	.44	Fire
62	161	223	49.7	.9	1.10	1.26		.52	PlaneJourneys
41	255	223	41.5	.9	1.09	1.35		.48	Weapons
BETTER FITTING OMITTED									
99	143	223	51.4	1.0	.89	.84	t	.61	Police
7	87	223	57.9	1.1	.88	.83	s	.56	CrossingStreets
46	81	223	58.8	1.2	.88	.80	r	.55	UglyPeople
30	236	223	43.1	.9	.87	.92	q	.49	SuddenNoise
54	266	223	40.5	.9	.87	1.58	p	.52	AngryPeople
75	285	223	38.8	.9	.85	.91	o	.55	MakingMistakes
97	65	222	61.3	1.3	.84	.73	n	.59	Kissing
61	313	223	36.2	1.0	.84	.88	m	.56	FeelingRejected
51	131	223	52.7	1.0	.84	1.03	l	.53	BeingTouched
64	298	223	37.6	.9	.83	1.05	k	.54	Disapproval
57	157	222	49.9	.9	.82	.77	j	.62	AnimalBlood
89	215	222	44.8	.9	.82	.80	i	.60	BeingSeenNude
40	149	223	50.8	.9	.82	.78	h	.64	DeadAnimals
36	233	223	43.4	.9	.82	.86	g	.55	ToughPeople
74	125	222	53.3	1.0	.79	.72	f	.58	Cripples
15	110	221	54.8	1.0	.79	.71	e	.62	Sirens
48	192	223	46.9	.9	.75	.75	d	.60	SickPeople
9	212	223	45.2	.9	.74	.78	c	.54	StrangePlaces
23	180	223	48.0	.9	.73	.77	b	.60	Strangers
68	130	223	52.8	1.0	.72	.75	a	.63	Darkness
MEAN	167	223	50.0	1.0	1.01	1.05			
SD	77	1	8.2	.2	.14	.22			

Note: WINSTEPS v3.08 Table 10.1 item statistics infit order output (Linacre, 2003).
Input: 223 persons, 108 fears analyzed, 3 categories.
CD = plot code.

Table 15 on page 69 provides category options and distractor frequencies by infit order. These additional details can be useful in determining how categories are working for each item. Entries 93A (Homosexuality), 92B (God's punishment), and 109C (Marriage) show category disorder among categories 2, 3, and 4. The counts and percentages, however, show that only a few persons produced these disruptions. Of course, the effect of these disruptions is removed by rescoring categories 2, 3, and 4 as all 2s.

Item 18E (High places), 19G (Look down from on high) and 6I (Speaking in public) also show a discontinuity. Although small, it can be observed in the 0–1 step shown in Table 15.

The category/option/distractor frequencies given by infit order in are useful for identifying the details of response irregularities. We must decide whether to tolerate these response irregularities because of their modest size or to make changes to the survey or the scoring model.

Table 15. *Fear Survey Category Option / Distractor Frequencies Infit Order*

NUM	DATA CODE	SCORE VALUE	DATA COUNT	%	USED COUNT	%	AVERAGE MEASURE	OUTFIT MNSQ	FEARS
93	A 0	0	149	67	149	67	41.93	1.2	Homosexuality
	1	1	31	14	31	14	46.51	1.0	
	4	2	14	6	14	6	48.31	1.9	
	2	2	15	7	15	7	49.94	1.7	
	3	2	12	5	12	5	56.18	1.6	
	missing	***	2	1	2	1	58.31		
92	B 0	0	119	53	119	53	41.32	1.2	GodsPunishment
	1	1	49	22	49	22	45.96	1.0	
	2	2	21	9	21	9	48.04	1.8	
	4	2	17	8	17	8	49.70	1.8	
	3	2	17	8	17	8	51.90	1.7	
104	C 0	0	126	57	126	57	41.10	1.2	Marriage
	1	1	38	17	38	17	45.92	1.1	
	2	2	31	14	31	14	46.49	2.1	
	4	2	11	5	11	5	53.49	.9	
	3	2	17	8	17	8	55.99	.8	
13	D 0	0	65	29	65	29	39.03	1.3	Dentists
	1	1	59	26	59	26	44.46	.6	
	3	2	32	14	32	14	46.40	2.0	
	2	2	48	22	48	22	47.36	1.5	
	4	2	19	9	19	9	51.93	.9	
18	E 0	0	108	49	108	49	40.66	1.1	HighPlaces
	1	1	48	22	48	22	44.48	1.8	
	2	2	32	14	32	14	48.68	1.6	
	4	2	17	8	17	8	50.75	1.2	
	3	2	17	8	17	8	53.71	.9	
	missing	***	1	0	1	0	44.90		
5	F 0	0	62	28	62	28	38.97	1.2	DeadPeople
	1	1	55	25	55	25	43.00	.7	
	2	2	59	27	59	27	47.70	1.4	
	3	2	25	11	25	11	48.77	1.2	
	4	2	21	9	21	9	49.82	.9	
	missing	***	1	0	1	0	44.90		
19	G 0	0	63	28	63	28	39.53	1.3	LookDownFromHigh
	1	1	68	30	68	30	43.88	1.5	
	3	2	28	13	28	13	46.99	1.3	
	2	2	41	18	41	18	47.43	1.6	
	4	2	23	10	23	10	50.90	.8	
4	H 0	0	101	45	101	45	40.74	1.3	LoudVoices
	1	1	60	27	60	27	44.23	1.1	
	4	2	9	4	9	4	48.71	1.4	
	2	2	36	16	36	16	49.84	1.2	
	3	2	17	8	17	8	53.22	1.1	
6	I 0	0	40	18	40	18	38.52	1.6	SpeakInPublic
	1	1	60	27	60	27	43.02	2.5	
	2	2	54	24	54	24	45.55	1.3	
	4	2	32	14	32	14	47.54	1.1	
	3	2	37	17	37	17	48.70	1.0	

Note: WINSTEPS v3.08 Table 10.3 category/option/distractor frequencies infit order output (Linacre, 2003). Input: 223 persons, 108 fears analyzed, 3 categories.

The Guttman diagram in Table 16 helps make these decisions.

Table 16. *Fear Survey Most Unexpected Responses Guttman Diagram*

```
NUM   FEAR          MEASURE                              PERSON
                                    11112    11   21 21111 11   11 2111121 2    2 1 11
                                    5555527581 878993059219799916804919967008811 275670
                                    6548766123860128054924549366537412839286271511740 9
                               high-------------------------------------------------------
    16 Failure        30.2        .........0..0..........00...00.. ................
    49 BeingCriticiz  36.7        .........1.0............................................
     6 SpeakingInPub  36.9 I      .1..1...0.1.............................................
    76 LookingFoolis  37.4 z      1.....0................................................
    64 Disapproval    37.6 k      .1..1....1.............................................
    47 Fire           40.0 Z      .1....1.1...0.........................2....2....
    54 AngryPeople    40.5 p      1101.....1............................2.............
     2 OpenWounds     41.4 V      .......11..............................2.....2.
    41 Weapons        41.5        1..11.0.1.............................2.............
    45 SeeingAFight   41.6        1.....0.1...0..........................................
    10 Falling        41.7 P      .1...................................2......2.....1...
    35 ToSeeBullying  41.9        1...1................................2.................
    55 Mice&Rats      42.5        .......0............................................2....
    24 Bats           43.6 R      ......00...0........................................2....
    94 WrongClothes   44.7        ..0. ......................2...2........2....
    22 Injections     44.8 W      ....10...............2.2...2..........2....
    27 Authorities    46.7        ...0.1...0.0.........2.....2..........2...2....
    69 MissedHeartBe  46.8        ......................22.....2.........2.....1..1
    18 HighPlaces     47.9 E      ...................2.......................1..
    62 PlaneJourneys  49.7        .......0...................2...........2...
     3 BeingAlone     49.8 L      ..........0...........2...................2.
   108 Hospitals      50.1 y      .1.....0..........2.........2.................1
    33 Crowds         50.6        ...........0..........2...........2.....2....
   105 Insecticides   52.1 U      .......0.....................2....2.....
    63 MedicalOdors   52.5        ...................2...........2.............1
    51 BeingTouched   52.7 l      1..............................22.1.2.1....
    38 DeepWater      52.8 N      1.....................2.....2.2.2.........
    93 Homosexuality  54.3 A      .....00.......2.....2.....2...2....2..1.......
    52 Elevators      55.2        .............2...............1..2......1.....
    85 OppositeSex    55.8        .......0.....2.................12........2....
    70 NudeMen        56.1 S      .......0....2........2.  .1.1..2.2...........
    90 Medicine       56.2        ...0....................2.........2........1..
    82 SightOfKnives  56.7        .............2....................2.1.......
   102 LeavingHome    57.9        ...........2............1....1..2..........
    91 SexualArousal  58.6 w      .......................1........2.........1..
    46 UglyPeople     58.8 r      .........2......21.....1.............
    21 ImaginaryCrea  58.9        ...............2.....2...21...............
    71 NudeWomen      59.3        ......................2.................
    11 Autos          59.4        ................2...22.1........
   101 Masturbation   60.4 Q      .. .....  ............2.......2.................
    43 BusJourneys    60.8        ...........2.........1...2.............
    97 Kissing        61.3 n      .............2................1.........
    95 Ministers&Pri  62.8        .................2....................
    34 Cats           63.6 O      .........2......2.......1............
    25 TrainJourneys  63.7        .....2........................1......
    31 CarJourneys    64.0        .........2......22................
    50 StrangeShapes  64.5        ...............2.........2..............1..
   100 Fish           67.2        ......................2....1............
    86 OpenSpaces     68.0        ...........22.......2..................
     1 NoiseOfVacuum  68.3        ...........1.2.............................
                               -------------------------------------------------------low
                                    5555511112887119321921119116811921111218215271611
                                  \ 6548727581 60898005921979391530411996700211 215470
                                    66123    12   54 24549 66   74 2839286 7    1 7 09
```

Note: WINSTEPS Table 10.5 most unexpected responses output (Linacre, 2003).
Input: 223 persons, 108 fears analyzed, 3 categories.

This Guttman diagram was produced using the three categories that were previously demonstrated as optimal for analysis of this instrument. The diagram lists each misfitting item and measure with a profile of unexpected responses. The most unexpected responses are identified so that patterns can be readily detected.

Figure 14 on page 72 and Figure 15 on page 73 are key maps of the 3-category fear yardstick with items listed in fear order along the right vertical axis and the person distribution recorded along the bottom of each page. They show the line of the conjoint response of persons to items. We see many stacks indicating measure-similar items and can identify the items associated with them. These stacks can help us to shorten the fear scale or to divide the scale into parallel surveys (i.e., different but shorter surveys of the same expected item response).

```
0    20   30   40   50   60   70   80   90  100   NUM   FEAR
|--//--+-----+-----+-----+-----+-----+-----+-----+-----|
                         0   :   1   :   2          1   NoiseOfVacuum
                         0   :   1   :   2         86   OpenSpaces
                       0   :   1   :   2          100   Fish
                       0   :   1   :   2           42   -Dirt
                      0   :   1   :   2            37   Birds
                      0   :   1   :   2            50   StrangeShapes
                     0   :   1   :   2             31   CarJourneys
                     0   :   1   :   2             32   DullWeather
                    0   :   1   :   2              25   TrainJourneys
                    0   :   1   :   2              34   Cats
                    0   :   1   :   2              95   Ministers&Priests
                   0   :   1   :   2               97   Kissing
                  0   :   1   :   2                43   -BusJourneys
                 0   :   1   :   2                101   Masturbation
                 0   :   1   :   2                 11   Autos
                 0   :   1   :   2                 71   NudeWomen
                0   :   1   :   2                  21   ImaginaryCreatures
                0   :   1   :   2                  46   UglyPeople
                0   :   1   :   2                  91   SexualArousal
                0   :   1   :   2                   7   CrossingStreets
                0   :   1   :   2                 102   LeavingHome
                0   :   1   :   2                  87   -Dogs
               0   :   1   :   2                   82   SightOfKnives
               0   :   1   :   2                   20   Worms
               0   :   1   :   2                   90   Medicine
               0   :   1   :   2                   70   NudeMen
              0   :   1   :   2                    85   OppositeSex
             0   :   1   :   2                     98   -Undertakers
             0   :   1   :   2                     52   Elevators
             0   :   1   :   2                     15   Sirens
             0   :   1   :   2                     93   Homosexuality
            0   :   1   :   2                      66   -Cemeteries
            0   :   1   :   2                      14   -Thunder
            0   :   1   :   2                      74   Cripples
            0   :   1   :   2                      38   DeepWater
            0   :   1   :   2                      68   Darkness
            0   :   1   :   2                      51   BeingTouched
            0   :   1   :   2                      63   MedicalOdors
           0   :   1   :   2                      105   Insecticides
           0   :   1   :   2                       72   -Lightening
           0   :   1   :   2                       78   Fainting
           0   :   1   :   2                       99   Police
           0   :   1   :   2                       73   Doctors
           0   :   1   :   2                       88   Germs
           0   :   1   :   2                       29   ToSeeInjections
           0   :   1   :   2                       40   DeadAnimals
          0   :   1   :   2                        33   Crowds
          0   :   1   :   2                       104   Marriage
          0   :   1   :   2                       108   Hospitals
          0   :   1   :   2                        80   -HarmlessSpiders
          0   :   1   :   2                        57   -AnimalBlood
          0   :   1   :   2                        92   GodsPunishment
          0   :   1   :   2                         3   BeingAlone
          0   :   1   :   2                        62   PlaneJourneys
          0   :   1   :   2                       107   BeingInCharge
|--//--+-----+-----+-----+-----+-----+-----+-----+-----|
0    20   30   40   50   60   70   80   90  100
              11 12112211
      21   1  18338315503066954641121 1111              PERSONS
              T   S    M     S    T
    MOST FEAR          MEAN         LEAST FEAR
```

Note: WINSTEPS Table 2.2 expected score output (Linacre, 2003).
Input: 223 persons, 108 fears analyzed, 3 categories.
M = mean. S = 1 satndard deviation. T = 2 standard deviations. ":" = half-score point

Figure 14. *Fear Survey Key Map (55 Least Fears)*

```
0    20   30   40   50   60   70   80   90  100   NUM   FEAR
|--//--+----+----+----+----+----+----+----+----|
              0   :   1   :   2                  106   Vomiting
              0   :   1   :   2                   79   Nausea
              0   :   1   :   2                   59   Enclosures
              0   :   1   :   2                   81   MakingDecisions
              0   :   1   :   2                  103   PhysicalExams
              0   :   1   :   2                   28   FlyingInsects
              0   :   1   :   2                   12   +BeingTeased
              0   :   1   :   2                   44   -CrawlingInsects
              0   :   1   :   2                   23   Strangers
              0   :   1   :   2                   18   HighPlaces
              0   :   1   :   2                   56   -HumanBlood
              0   :   1   :  2                     4   +LoudVoices
              0   :   1   :  2                    17   EnterToOthrsSeated
             0   :   1   :   2                    65   -HarmlessSnakes
           0   :   1   :   2                      48   SickPeople
           0   :   1   :   2                      69   MissedHeartBeats
           0   :   1   :   2                      27   +Authorities
           0   :   1   :   2                      39   WatchedAtWork
           0   :   1   :   2                      83   MentalIllness
           0   :   1   :   2                      26   +FeelingAngry
          0   :   1   :   2                        9   StrangePlaces
          0   :   1   :   2                       58   PartingFriends
          0   :   1   :   2                       22   Injections
          0   :   1   :   2                       89   BeingSeenNude
          0   :   1   :   2                       94   WrongClothes
          0   :   1   :  2                         67   +BeingIgnored
         0   :   1   :   2                         24   Bats
         0   :   1   :   2                         36   ToughPeople
         0   :   1   :   2                         30   SuddenNoise
         0   :   1   :   2                         84   TakingTests
         0   :   1   :  2                           55   Mice&Rats
         0   :   1   :   2                          53   SeeingSurgery
        0   :   1   :   2                           35   ToSeeBullying
        0   :   1   :   2                           10   Falling
        0   :   1   :   2                           19   LookingDwnFromHigh
        0   :   1   :   2                           45   SeeingAFight
        0   :   1   :   2                           41   Weapons
        0   :   1   :   2                            2   OpenWounds
        0   :   1   :   2                           13   Dentists
        0   :   1   :   2                            8   InsanePeople
       0   :   1   :   2                            54   +AngryPeople
       0   :   1   :   2                             5   DeadPeople
       0   :   1   :   2                            47   Fire
      0   :   1   :   2                             75   +MakingMistakes
      0   :   1   :   2                             77   LosingControl
      0   :   1   :  2                              64   +Disapproval
      0   :   1   :   2                             76   +LookingFoolish
     0   :   1   :   2                               6   SpeakingInPublic
     0   :   1   :   2                              60   SurgeryProspects
     0   :   1   :   2                              49   +BeingCriticized
     0   :   1   :   2                              96   HurtingFeelings
     0   :   1   :   2                              61   +FeelingRejected
    0   :   1   :   2                               16   +Failure
|--//--+----+----+----+----+----+----+----+----|
0    20   30   40   50   60   70   80   90  100
                11 12112211
       21   1  18338315503066972641121l 111                 PERSONS
             T    S    M    S    T
    MOST FEAR              MEAN              LEAST FEAR
```

Note: WINSTEPS Table 2.2 expected score output.
Input: 223 persons, 108 fears analyzed, 3 categories.
M = mean. S = 1 satndard deviation. T = 2 standard deviations. ":" = half-score point

Figure 15. *Fear Survey Key Map (53 Greatest Fears)*

Individual person key maps in Figure 16 and Figure 17 on page 75 and Figure 18 on page 76 provide examples taken from the sample of persons who responded to the fear scale. In these examples only the items that expose the misfit are listed on the right side.

```
PER      NAME       MEASURE   INFIT  (MNSQ)  OUTFIT   SE
71     07114600       31.0     1.5     T       1.4    .22    NUM   FEAR
0     20   30    40    50    60    70    80    90   100
|--//--+-----+-----+-----+-----+-----+-----+-----+-----|
              0                     (2)                     85   OppositeSex
              0                (2)                           33   Crowds
              0              (2)                             17   EnterToOthrsSeatd
              0              (2)                             27   Authorities
              0        .1.                                   39   WatchedAtWork
              0        .1.                                   83   MentalIllness
              0              (2)                             22   Injections
              0        .1.                                   84   TakingTests
              0        .1.                                   55   Mice&Rats
              0           (2)                                47   Fire
          .0. 1                                             75   MakingMistakes
          .0. 1                                             64   Disapproval
              1      .2.                                     76   LookingFoolish
              1      .2.                                      6   SpeakingInPublic
          .0. 1                                             60   SurgeryProspects
              1      .2.                                     96   HurtingFeelings
          .0. 1                                             61   FeelingRejected
          .0. 1                                             16   Failure
```

```
PER      NAME       MEASURE   INFIT  (MNSQ)  OUTFIT   SE
50     07114600       32.5     1.4     Z       .9     1.8    NUM   FEAR
0     20   30    40    50    60    70    80    90   100
|--//--+-----+-----+-----+-----+-----+-----+-----+-----|
              0                 (2)                         59   Enclosures
              0        .1.                                  69   MissedHeartBeats
              0                (2)                          83   MentalIllness
              0                (2)                          94   WrongClothes
              0                (2)                          67   BeingIgnored
              0        .1.                                  84   TakingTests
              0                (2)                          19   LookingDwnFrmHigh
              0                (2)                          13   Dentists
          .0. 1                                             54   AngryPeople
          .0. 1                                              5   DeadPeople
          .0. 1                                             47   Fire
              1      .2.                                     75   MakingMistakes
              1      .2.                                     77   LosingControl
              1      .2.                                     64   Disapproval
              1      .2.                                     76   LookingFoolish
          .0. 1                                              6   SpeakingInPublic
          .0. 1                                             60   SurgeryProspects
          .0. 1                                             49   BeingCriticized
              1      .2.                                     96   HurtingFeelings
              1      .2.                                     61   FeelingRejected
              1    .2.                                       16   Failure
```

Note: WINSTEPS v3.08 diagnostic key forms for individuals output.
Input: 223 persons, 108 fears analyzed, 3 categories v3.08.
MNSQ letter is on fit plots.

Figure 16. *Fear Survey Person Key Maps (Persons 71 and 50)*

```
PER    NAME      MEASURE   INFIT   (MNSQ)   OUTFIT   SE
157    15713900   22.8      1.3      NR       1.3     2.9    NUM   FEAR

0     20    30    40    50    60    70    80    90    100
|--//--+-----+-----+-----+-----+-----+-----+-----+-----|
           0                (1)                              14   Thunder
           0                (1)                              72   Lightening
           0                        (2)                      62   PlaneJourneys
           0            (1)                                  69   MissedHeartBeats
           0          (1)                                    10   Falling
           0                (2)                              13   Dentists
           0        .1.                                      47   Fire
           0      .1.                                        60   SurgeryProspects
      .0.  1                                                 16   Failure
```

```
PER    NAME      MEASURE   INFIT   (MNSQ)   OUTFIT   SE
55     05511800   72.9      1.0      NR       2.2     2.3    NUM   FEAR

0     20    30    40    50    60    70    80    90    100
|--//--+-----+-----+-----+-----+-----+-----+-----+-----|
                                  1    .2.                    1   NoiseOfVacuum
                                  1    .2.                   86   OpenSpaces
                                  1    .2.                  100   Fish
                              .1.      2                     31   CarJourneys
                            .1.        2                     11   Autos
                          .1.          2                     85   OppositeSex
                          .1.          2                     52   Elevators
                        (1)            2                     68   Darkness
                        (1)            2                     40   DeadAnimals
                        (1)            2                    108   Hospitals
                      (1)              2                     23   Strangers
                      (1)              2                     26   FeelingAngry
                      (1)              2                     58   PartingFriends
                    (1)                2                     10   Falling
                    (1)                2                     54   AngryPeople
                    (1)                2                     47   Fire
                    (1)                2                     64   Disapproval
                    (1)                2                      6   SpeakingInPublic
```

```
PER    NAME      MEASURE   INFIT   (MNSQ)   OUTFIT   SE
64     06411700   16.5      1.0      NR       3.1     4.1    NUM   FEAR

0     20    30    40    50    60    70    80    90    100
|--//--+-----+-----+-----+-----+-----+-----+-----+-----|
           0                    (1)                          50   StrangeShapes
           0              (1)                                91   SexualArousal
           0              (1)                                90   Medicine
           0          (1)                                    18   HighPlaces
           0        (1)                                      19   LookingDwnFrmHigh
```

Note: WINSTEPS v3.08 7.2 diagnostic key forms for individuals output.
Input: 223 persons, 108 fears analyzed, 3 categories v3.08.
NR = non reported.

Figure 17. *Fear Survey Person Key Maps (Persons 157, 55, and 64)*

```
PER    NAME      MEASURE   INFIT   (MNSQ)   OUTFIT   SE
56     05611700  75.2      1.0      NR       2.0     2.0    NUM   FEAR

0    20   30   40   50   60   70   80   90  100
|--//--+-----+-----+-----+-----+-----+-----+-----+-----|
                              1   .2.            1    NoiseOfVacuum
                              1   .2.           86    OpenSpaces
                              1   .2.          100    Fish
                        .1.   2                 32    DullWeather
                         .1.  2                 97    Kissing
                        .1.   2                 11    Autos
                         .1.  2                102    LeavingHome
                         .1.  2                 87    Dogs
                   (1)        2                 38    DeepWater
                   (1)        2                 68    Darkness
                   (1)        2                 51    BeingTouched
              (1)            2                  35    ToSeeBullying
              (1)            2                  45    SeeingAFight
              (1)            2                  41    Weapons
             (1)            2                   54    AngryPeople
           (1)            2                     76    LookingFoolish
```

```
PER    NAME      MEASURE   INFIT   (MNSQ)   OUTFIT   SE
56     05611700  75.2      1.0      NR       2.0     2.0    NUM   FEAR

0    20   30   40   50   60   70   80   90  100
|--//--+-----+-----+-----+-----+-----+-----+-----+-----|
  0                       (2)                   3    BeingAlone
  0                 (2)                          2    OpenWounds
```

```
PER    NAME      MEASURE   INFIT   (MNSQ)   OUTFIT   SE
196    19604400  38.0      .8       NR       1.3     1.5    NUM   FEAR

0    20   30   40   50   60   70   80   90  100
|--//--+-----+-----+-----+-----+-----+-----+-----+-----|
           0                       (2)          100   Fish
           0    .1.                              72   Lightening
           0    .1.                              99   Police
           0    .1.                              29   ToSeeInjections
           0    .1.                              40   DeadAnimals
           0   .1.                               57   AnimalBlood
           0           (2)                      103   PhysicalExams
           0    .1.                              56   HumanBlood
           0    .1.                              17   EnterToOthersSeated
           0           (2)                        65   HarmlessSnakes
           0   .1.                               69   MissedHeartBeats
        .0.   1                                  26   FeelingAngry
        .0.   1                                   9   StrangePlaces
        .0.   1                                  58   PartingFriends
              1        (2)                       89   BeingSeenNude
         .0.  1                                  30   SuddenNoise
              1    .2.                           84   TakingTests
        .0.   1                                  35   ToSeeBullying
        .0.   1                                   5   DeadPeople
              1   .2.                            64   Disapproval
              1   .2.                             6   SpeakingInPublic
        .0.   1                                  60   SurgeryProspects
              1   .2.                            96   HurtingFeelings
```

Note: WINSTEPS v3.08 Table 7.2 diagnostic key forms for individuals output.
Input: 223 persons, 108 fears analyzed, 3 categories v3.08.
NR = nonreported.

Figure 18. *Fear Survey Person Key Maps (Persons 56, 170, and 196)*

Above, the indivual person keymaps include the person identification number, demographic code, their measure, infit, and outfit. We can make a tentative diagnosis from the misfitting responses exposed. For person 71, the 2s in the misfit profile suggest unusual shyness in personal inter- actions in various settings. The misfit profile for person 50, however, suggests unusual shame in a variety of settings. The misfit profile of person 157 implies agoraphobia. Finally, person 64 has a much lower measure, with the 1s suggesting specific settings that are especially fear inducing to person 64.

Examine the misfit profile of person 55 in Figure 17 on page 75. What does it suggest? What do profiles suggest of person 56, person 170, and person 196 in Figure 18 on page 76? The application of substantive theory, together with comprehensive measurement software provide power- ful tools to enhance our analysis and inform our judgements.

Having concluded an exposition and analysis of dicho- tomous and rating scale data in some detail, we now proceed with another example and application. It entails building rulers for measuring reader ability and text readability.

Chapter 9 Uniform Reading and Readability Measures

by A. Jackson Stenner and Ben Wright[1]

The world of education has long been waiting for a sunrise. Believe it or not, a popular compilation of educational tests lists 97 different reading tests (Mitchell, 1985). This situation produces 97 different "reading ability measures." What a mess! But now, with the dawn of uniform educational measures, the sun is rising.

Measures are older than talking. Birds measure. So do bees. Our own measures evolved from our bodies — our feet, our arms, our hands, our fingers. An inch is the distance from thumb tip to knuckle. A span is the distance between thumb tip and little finger. A cubit is the length of a forearm. A fathom is the distance between outstretched arms. A pace is two steps. A furlong is 200 paces. A mile is 1,000 paces.

Abstractly equal units of length were counted on before the oldest writing fragments. Figure 19 on page 80 is Moses' plan for the Tabernacle. Without approximations to equal units, Summerians, Babylonians, Egyptians and Hebrews could not have imagined, let alone built, their towers.

---

1. Adaptation of a presentation at the Association of Test Publishers Career Achievement Award in Computer-Based Testing for Benjamin Drake Wright, Ph.D, San Diego, February, 2002, based on "Measuring Reading,"a paper presented by Wright and Stenner at the International Seminar on Developmental Assessment, Melbourne, Australia, July, 1998.

> 2. THE LENGTH OF ONE CURTAIN SHALL
> BE EIGHT AND TWENTY CUBITS, AND
> THE BREADTH OF ONE CURTAIN FOUR
> CUBITS; AND EVERY ONE OF THE
> CURTAINS SHALL HAVE ONE MEASURE.

Figure 19. *Exodus 26*

Fair measurement is embedded in Judeochristian morality. But the "perfect and just measure" demanded in Deuteronomy 25 (Figure 20) is an ideal that can only be approximated in practice. The "weight" referred to is a shekel stone, which was understood to weigh 11.4 ounces. However, archeologists have never found two shekel stones that weighed exactly the same. No technology, no matter how advanced, can fabricate perfect weights. Nevertheless, even when Deuteronomy was written, we already understood the essential necessity and justice of fair units.

> 13.THOU SHALT NOT HAVE IN THY BAG
> DIVERSE WEIGHTS, A GREAT AND A
> SMALL.
> 14. THOU SHALT NOT HAVE IN THINE
> HOUSE DIVERSE MEASURES, A GREAT
> AND A SMALL.
> 15. THOU SHALT HAVE A PERFECT AND
> JUST WEIGHT, A PERFECT AND JUST
> MEASURE.

Figure 20. *Deuteronomy 25*

A fair weight of seven was also a tenant of faith among seventh century Muslims. Muslim leaders were censured for using less "righteous" standards (Sears, 1997). In Figure 21, we see that 12 centuries ago Caliph 'Umar b. 'Abd al-'Aziz ruled

> THE PEOPLE OF AL-KUFA HAVE BEEN STRUCK WITH TRIAL, HARDSHIP, OPPRESSIVE GOVERNMENTS AND WICKED PRACTICES. THE RIGHTEOUS LAW IS JUSTICE AND GOOD CONDUCT. I ORDER YOU TO TAKE IN TAXES ONLY THE WEIGHT OF SEVEN.

Figure 21. *Damascus, 723*

The impetus for uniformity in the representation of quantity appears again in King John's Magna Carta (see Figure 22). Without the ideal of uniform measures, there would be no money. There would be no fitted clothes, because there would be no way to fit them. Imagine what life would be like if there were no abstract unit of length such as the inch. Imagine that an inch is complex — and differs with every situation and material. Imagine that wood inches are different from brick inches, that those are different from steel inches. We would not have civilization. We would have a mess — a mess like the mess that permeates most of what we misleadingly refer to as "educational tests and measurements."

> 35. THERE IS TO BE ONE MEASURE
> OF WINE AND ALE AND CORN
> WITHIN THE REALM, NAMELY THE
> LONDON QUARTER, AND ONE
> BREADTH OF CLOTH, AND IT SHALL
> BE THE SAME WITH WEIGHTS.

Figure 22. *The Magna Carta, Runnymede, 1215*

The Evolution of Science

The study of any subject begins with tangles of speculations. Ideas branch in all directions. As we work through the tangle, we connect what we experience with what we see. We coax our ideas into shape, form unities, and develop lines of inquiry. We fit our ideas together and make them into something. We evolve our bush of ideas into a tree of knowledge. The bush was a tangle. The tree has direction. Our final step in wrestling a useful abstract assertion from a complex concrete confusion is to carve a ruler out of our tree. The ruler does not exist until we imagine it and carve it. The carving is not perfect. It is just an approximation. But what it approximates — a perfectly straight line — enables us to use it as though it was marked off in perfectly equal intervals.

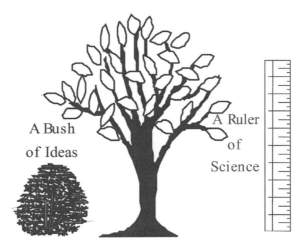

Figure 23. *A Tree of Knowledge*

We can pace off land in somewhat equal steps, but steps inevitably vary according to conditions. To produce reliable measurements, we need something more reproducible than pacing. The scientific measurement of length was born as we connected our experience of stride with manmade marks on straight pieces of wood extracted from tree trunks. A piece of tree is more stable than any individual person's paces. A ruler does not change its benchmarks. When we grow a confusing bush of tangled ideas into a tree of useful knowledge and make a ruler, then we can plan and build a pyramid, a temple, a house — and also measure the height of a child.

The Imaginary Inch

An *inch* is pure, abstract, and without content. It has no meaning of its own. It is an imaginary unit of length. A height of inches, however, has meaning. As we grow, we learn the advantages to growing taller. Brick size has meaning. As we build, we learn the advantages of same-sized bricks. What makes bricks useful is that their interchangeability is maintained by approximations to the fiction we call an inch.

It is essential that our idea of an abstract inch is always the same. If we let our idea of an inch change each time we make a measure, we cannot produce useful bricks or keep track of our child's growth. As our child grew, we would not know by how much they had grown. But with a uniform unit of measurement, such as an inch, when we measure the height of our children, we can refer to last year — or perhaps to the height of an average second grader because, as it turns out, even though school has no effect on height, child height is related to school grade. Figure 24 shows how we can guess a child's grade by how tall they are — and the height of the child by the grade. That is an understanding based entirely on applications of rulers. The applications would be useless without that single, unvarying inch that our rulers approximate.

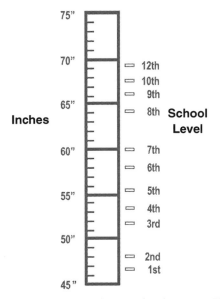

Figure 24. *Educational Status by Average Height*

No metric has content of its own. The ruler, with its equal measurement units, is merely an approximate realization of a pure idea — an ideal that we invented from tangled experiences of length. We invented the ruler as a device by which to make uniform measures available for any application we may care to undertake.

One Kind of Reading Ability

Let's turn to the measurement of reading. We can think of reading as the tree in Figure 25. It has roots such as oral comprehension and phonological awareness. As reading ability grows, a trunk extends through grade school, high school, and college and branches at the top into specialized vocabularies. That single trunk is longer than many realize. It grows quite straight and singular from first grade through college.

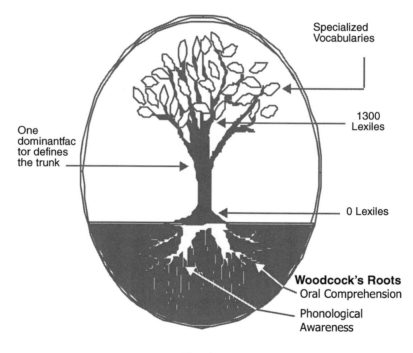

Figure 25. *The Reading Tree*

Reading has always been the most-researched topic in education (Thorndike & Hagen, 1965). There have been many studies of reading ability, large and small, local and national. When we review the results of these studies, one clear picture emerges. Despite the 97 ways to test reading ability, many decades of empirical data document definitively that no researcher has been able to measure more than one kind of reading ability. This has proven true in spite of intense interest in discovering diversity. Consider three examples: the 1940s *Davis Study*, the 1970s *Anchor Study* and six 1980s and 1990s studies by the Educational Testing Service (ETS).

Davis (1940s). Fred Davis went to a great deal of trouble to define and operationalize nine kinds of reading ability (1944). He made up nine different reading tests to prove the separate identities of his nine kinds. He gave his nine tests to hundreds of students, analyzed their responses to prove his thesis, and reported that he had established nine kinds of reading. But when Louis Thurstone (1946) reanalyzed Davis' data, Thurstone showed conclusively that Davis had no evidence of more than one dimension of reading.

Anchor Study (1970s). In the 1970s, worry about national literacy prompted the U.S. government to finance a national *Anchor Study* (Jaeger, 1973). Fourteen different reading tests were administered to a great many children to uncover the relationships among the 14 different test scores. Millions of dollars were spent. Thousands of responses were analyzed. The final report required 15,000 pages in 30 volumes — just the kind of document one reads overnight, takes to school the next day, and applies to teaching (Loret, Seder, Bianchini, & Vale, 1974). In reaction to this futility, and against a great deal of proprietary resistance, Rentz and Bashaw (1975, 1977) were able to obtain a small grant to reanalyze the *Anchor Study* data. By applying new methods for constructing objective measurement (Wright & Stone, 1979), Rentz and Bashaw were able to show that all 14 tests used in the *Anchor Study* — with all their different kinds of items, item authors, and publishers — could all be calibrated onto one linear *National Reference Scale* of reading ability.

The essence of Rentz and Bashaw's (1977) results can be summarized on one easy-to-read page — a bit more useful than 15,000 pages. Their one-page summary shows how every raw score from the 14 *Anchor Study* reading tests can be equated to one linear *National Reference Scale*. Their page also shows that the scores of all 14 tests can be understood as measuring the same kind of reading on one common scale. The Rentz and Bashaw *National Reference Scale* is additional evidence that, so far, no more than one kind of reading ability has ever been measured. Unfortunately, their work had little effect on the course of U.S. education. The experts went right on claiming that there must be more than one kind of reading — and sending teachers confusing messages as to what they were supposed to teach and how to do it.

ETS Studies (1980s and 1990s). In the 1980s and 1990s, the
ETS did a series of studies for the U.S. government. ETS
(1990) insisted on three kinds of reading: *prose* reading, *document* reading, and *quantitative* reading. They built a separate test to measure each of these three kinds of reading,
greatly increasing costs. Versions of these tests were administered to samples of school children, prisoners, young
adults, mature adults, and senior citizens. ETS reported
three reading measures for each person and claimed to have
measured three kinds of reading (Kirsch & Jungeblut,
1986). But reviewers noted that no matter which kind of
reading was chosen, there were no differences in the results
(Reder, 1996; Zwick, 1987). When the relationships among
reading and age and ethnicity were analyzed, whether for
prose, document, or quantitative reading, all conclusions
were the same.

Later, when the various sets of ETS data were reanalyzed by independent researchers, no evidence for three
kinds of reading measures could be found (Bernstein &
Teng, 1989; Reder, 1996; Rock & Yamamoto, 1994; Salganik
& Tal, 1989; Zwick, 1987). The correlations among ETS
prose, document, and quantitative reading measures ranged
from 0.89 to 0.96. Thus, once again and in spite of strong
proprietary and theoretical interests in proving otherwise,
nobody had succeeded in measuring more than one kind of
reading ability.

Lexiles

Figure 26 on page 91 is a reading ruler. Its Lexile units work just like the inches in Figure 24 on page 85. The Lexile ruler is built out of readability theory, school practice, and educational science. The Lexile scale is an interval scale. It comes from a theoretical specification of a readability unit that corresponds to the empirical calibrations of reading test items. It is a readability ruler. It is a reading ruler that conjointly measures reader ablity and text readability.

Readability formulas are built out of abstract characteristics of language. No attempt is made to identify what a word or sentence means. The idea is not new. In 400 B.C.E., the Athenian Bar Association used readability calculations to teach lawyers to write briefs (Chall, 1988; Zakaluk & Samuels, 1988). According to the Athenians, the ability to read a passage was not the ability to interpret what the passage was about. The ability to read was just the ability to read. In 700 B.C.E., Talmudic teachers who wanted to regularize their students studies used readability measures to divide the Torah readings into equal portions of reading difficulty (Lorge, 1939). Like the Athenians, their concern in doing this was not with what a particular Torah passage was about, but rather the extent to which passage readability burdened readers.

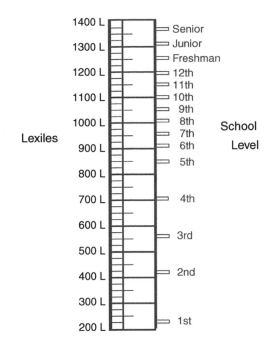

Figure 26. *Educational Status By Average Lexile*

In the 20th century, every imaginable structural characteristic of a passage has been tested as a potential source
for a readability measure: the number of letters and sylla-
bles in a word; the number of sentences in a passage; sentence length; balances between pronouns and nouns, verbs
and prepositions (Stenner, 1992, 1996). The Lexile readability measure uses word familiarity and sentence length.

Lexile Accuracies

Table 17 lists the correlations between readability measures
from the 10 most-studied readability equations and student
responses to different types of reading test items. The col-

umns of Table 17 report on five item types: Lexile slices, SRA passages, Battery Test sentences, Mastery Test cloze gaps, and Peabody Test pictures. The item types span the range of reading comprehension items. The numbers in the table show the correlations between theoretical readability measures of item text and empirical item calibrations calculated from students' test responses.

Table 17. *Empirical and Theoretical Item Difficulty Correlations*

Readability Equation	Test Item Type				
	Slice	Passage	Sentence	Cloze	Picture
Lexile	.90	.92	.74	.85	.94
Flesch	.85	.94	.70	.85	.85
ARI	.85	.93	.71	.85	.85
FOG	.85	.92	.75	.73	.85
Powers	.82	.93	.65	.83	.74
Holquist	.81	.91	.84	.81	.86
Flesch-1	.79	.92	.61	.81	.69
Flesch-2	.75	.87	.52	.70	.71
Coleman	.74	.87	.75	.75	.83
Dale-Chall	.76	.92	.73	.82	.67

Consider the top row. The Lexile readability equation predicted how difficult Lexile slices would be for persons taking a Lexile reading test at a correlation of 0.90, the SRA passage at 0.92, the Battery Sentence at 0.85, the Mastery Cloze at 0.74, and the Peabody Picture at 0.94 (Stenner, 1996). With the exception of the cloze items, these predictions are nearly perfect. Also note that the simple Lexile equation, based only on word familiarity and sentence length, predicted empirical item responses as well as any other readability equation — no matter how simple or complex the comparison. Table 17 documents yet again that one, and only one, kind of reading is measured by these reading tests. Were that not so, the array of nearly perfect correlations could not occur. Table 17 also shows that we can have a useful measurement of text readability and reader reading ability on a single reading ruler!

An important tool in reading education is the basal reader. The teaching sequence of basal readers records generations of practical experience with text readability and its bearing on student reading ability. Table 18 on page 94 lists the correlations between Lexile Readability and passage difficulty for the basal readers that are most used in the United States. Each series is built to mark out successive units of increasing reading difficulty.

Ginn has 53 units — from book 1 at the easiest to book 53 at the hardest. HBJ Eagle has 70 units. Teachers work their students through these series from start to finish. Table 18 shows that the correlations between Lexile measures of the texts of these basal readers and their sequential positions from easy to hard are extraordinarily high. In fact, when corrected for attenuation and range restriction, these correlations approach unity (Stenner, 1992, 1996).

Table 18. *Correlations Between Basal Reader Order and Lexile Readability*

Basal Reader Series	No. Units	Mean	(SD)	r	R	R'
Ginn	53	522	(272)	.93	.98	1.00
HBJ Eagle	70	549	(251)	.93	.98	1.00
SF Focus	92	552	(159)	.84	.99	1.00
Riverside	67	609	(231)	.87	.97	1.00
HM (1983)	33	449	(284)	.88	.96	.99
Economy	67	639	(265)	.86	.96	.99
SF American Tradition	88	696	(265)	.85	.97	.99
HBJ Odyssey	38	662	(209)	.79	.97	.99
Holt	54	615	(299)	.87	.96	.98
HM (1986)	46	707	(256)	.81	.95	.97
Open Court	52	694	(197)	.54	.94	.97

Note: Adapted from Stenner and Burdick (1997).

r = Pearson product-moment correlation; R = corrected for attenuation; R' = corrected for attenuation and range restriction.

All designers of a basal reader series have used their own ideas, consultants, and theory to decide what was easy and what was hard. Nevertheless, when the texts of these basal units are Lexiled, these Lexiles predict exactly where each book stands on its own reading ladder. This constitutes more evidence that despite differences among publishers and authors, all units end up benchmarking the same single dimension of reading ability.

Finally, there are the ubiquitous reading ability tests that are administered annually to assess every student's reading ability. Table 19 shows how well theoretical item text Lexiles predict item test performances on eight of the most popular reading tests. The second column shows how many passages from each test were Lexiled. The third column lists the item type. Once again there is a very high correlation between the difficulty of these items as calculated by the entirely abstract Lexile specification equation and the live data produced by students answering these items on reading tests. When we correct for attenuation and range restriction, the correlations are just about perfect.

Table 19. *Correlations Between Passage Difficulty and Lexile Readability*

Test	No. of Passages	Mean	(SD)	r	R	R'
SRA	46	644	(353)	.95	.97	1.00
CAT-E	74	789	(258)	.91	.95	.98
CAT-C	43	744	(238)	.83	.93	.96
CTBS	50	703	(271)	.74	.92	.95
NAEP	70	833	(263)	.65	.92	.94
Lexile	262	771	(463)	.93	.95	.97
PIAT	66	939	(451)	.93	.94	.97

Note: Adapted from Stenner and Burdick (1997).

r = Pearson product-moment correlation; R = corrected for attenuation; R' = corrected for attenuation and range restriction.

What does this mean? Not only is only one reading ability being measured by all of these reading comprehension tests, but we can replace all of the expensive data that are used to calibrate these tests empirically with one theoretical formula: the abstract Lexile specification equation. We can calibrate the reading difficulty of test items by Lexiling their text without administering them to a single student!

Figure 27 puts the relationship between theoretical Lexiles and observed item difficulties into perspective. The uncorrected correlation of 0.93, when disattentuated for error and corrected for range restrictions, approaches 1.00. The Lexile equation produces an almost perfect correlation between theory and practice.

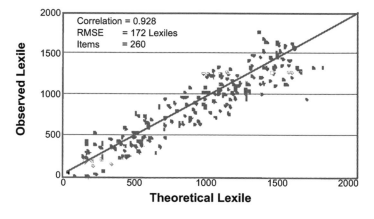

Figure 27. *Theory Into Practice*

Figure 27 shows the extent to which idiosyncratic variations in student responses and item response options enter the process. Where does this variation come from? Item response options have to compete with each other or they do not work. But there has to be one correct answer. Irregularity in the composition of multiple choice options, even when they are reduced to choosing one word to fill a blank, is unavoidable. What the item writers choose to ask about a passage and the options they offer the test taker are not only about reading ability. They are also about personal differences among test writers.

There are also variations among test takers in alertness and motivation that disturb their performances. In view of these unavoidable contingencies, it is surprising that the correlation between Lexile theory and actual practice is so high.

How does this affect the measurement of reading ability? The root mean square measurement error for a one-item test would be about 172 Lexiles. What are the implications of that much error? The distance from first-grade school books to second-grade school books is 200 Lexiles. So we would undoubtedly be uneasy with measurement errors as large as 172 Lexiles. However, when we combine the responses to a test of 25 Lexile items, the measurement error drops to 35 Lexiles. And when we use a test of 50 Lexile items, the measurement error drops to 25 Lexiles — one-eighth of the 200 Lexile difference between first- and second-grade books. Thus, when we combine a few Lexile items into a test, we get a measure of the reader's location on the Lexile reading ability ruler that is precise enough for most practical purposes. We do not plumb their depths of understanding, but we do measure their reading ability.

Lexile Items

One might now ask, how hard is it to write a Lexile test item? Figure 28 describes a study designed to find out whether Lexile items written by different authors produce usefully equivalent results (Stenner, 1998). Five apprentice item authors were each asked to choose their own text passages and to write their own response illustrated missing word options (see Figure 29 on page 99). Each author wrote 60 items that ranged from 900 to 1300 Lexiles. From these (5 x 60 = 300) items, five 60-item tests were constructed by drawing 12 items at random from each author. Then seven grade-school students each completed a different test each day for five days. This produced five measures for each student over the five days, and, by pooling days, five measures for each student over the five authors.

Step 1. 5 different authors compose 5 different sets of 60 Lexile items evenly sequenced from 900L to 1300L.

Step 2. 5 different 60-item tests are assembled. Each test constains 12 items selected at random from each author's set of 60 items.

Step 3. 7 students take a different 60-item test each day for 5 days.

Result. For each student, this produces:

- 5 measures across 5 days balanced over authors, and

- 5 measures across 5 authors balanced over days.

Figure 28. *Stability Study*

The question becomes, "Is the variation by author in a student's reading ability measure any larger than the variation by day?" If not, that would imply that writing useful Lexile test items, as in Figure 29, was not a problem, as even apprentice authors can do it well enough to obtain measures as stable as the differences in a person's reading performance from day to day.

Wilber likes Charlotte better and better each day. Her campaign against insects seemed sensible and useful. Hardly anybody around the farm had a good word to say for a fly. Flies spent their time pestering others. The cows hated them. The horses hated them. The sheep loathed them. Mr. and Mrs. Zuckerman were always complaining about them, and putting up screens. **Everyone _____ about them.**

a) agreed
b) gathered
c) laughed
d) learned

from *Charlotte's Web* by E. B. White (1952), New York; Harper and Row.

Figure 29. *An 800-Lexile Slice Test Item*

We know that each person's reading performance varies from day to day. Each performance depends on what is happening in our lives, what we have for breakfast, what happens at home, what happens at school, and how we feel about the test. Figure 30 on page 100 shows the day-to-day results for Emily and Randall. The vertical bars mark a 75% confidence region for the reading ability measure on each day. The up and down movements of the bars show how much these estimates of reading ability changed from day to day. On Monday, Randall and Emily did relatively well. On Tuesday, their performances sank. On Wednesday, they came back. On Thursday, Emily went up, but Randall went down. Finally, on Friday, they both went down. Figure 30 shows the differences a day makes in the reading performance of these two students. It reminds us that when we talk about reading ability, we must remember that performances vary from day to day.

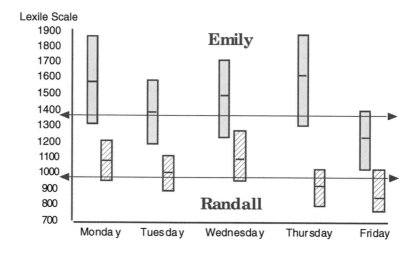

Figure 30. *Reading Ability Instability by Day*

Figure 31 on page 101 shows the variation in reading measures by item author. Notice that the variation among item authors in Figure 31 is no greater than the variation over days in Figure 30. No more noise is introduced into the

Lexile way of making a reading measure by a difference among item authors than by the difference a day makes.

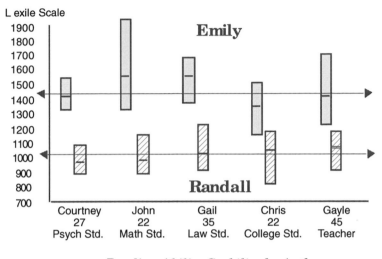

Figure 31. *Reading Ability Stability by Author*

These five Lexile item authors were not experts. They were just well-educated persons who received four hours of instruction in Lexile item writing. Courtney, 27, is a psychology student. John, 23, is a math student. Gail, 35, is law student. Chris, 22, is a football player. Gayle, 45, is a teacher.

Calculating Lexiles

Lexile measures of reading (Figure 32) are easy to understand and easy to use. Lexile readability, measured by word familiarity and sentence length, establishes how difficult a text is to read. Lexile reading ability, measured by how well a reader is able to recognize words and connect them into sentences, establishes a reader's ability to read a text.

Readability is passage reading difficulty measured in Lexiles.

Reading ability is ability to read passages measured in Lexiles.

Comprehension is defined as the difference between reader ability and text readability.

To measure Lexile reading ability, find out what Lexile passage readability a person can read with 75% success.

The Lexile formula is based on two axioms.

- The *semantic axiom*: The more familiar the words, the easier the passage is to read; the more unfamiliar the words, the harder.

- The *syntactic axiom*: The shorter the sentences, the easier the passage is to read; the longer the sentences, the harder.

Figure 32. *Lexile Measures of Reading*

The axioms in Figure 32 apply to whatever is read, quite apart from content. They apply whether we like what we are reading or not, whether it is prose, document, or quantitative material.

The Lexile system calculates passage readability from just these two characteristics — both of which are explicit in the passage. Sentence lengths are there to see. We count and average them. Word familiarities are obtained from compilations of word usage. The Lexile Analyzer originally used Carroll, Davies, and Richmond's sample of five million words (*Word Frequency Book,* 1971).[2] It now embodies a corpus of 500 million words used to compute word frequencies.

Step 1. Divide the book into natural slices of 125–140 words.

Step 2. For each slice i, determine

- log mean sentence length (SL_i) and

- mean log word frequency (WF).

Step 3. Calibrate the Lexile measure of slice i using the equation:

$$\text{Readability} = 582 + 1768 S_{Li} - 386 WF_i \qquad (23)$$

The Lexile measure of a book is equal to the Lexile level of a reader who succeeds on 75% of that book's slices.

Figure 33. *How to Calculate a Lexile Book Measure*

If readers do not know the words, they cannot read the passage. If they do know the words, they can begin to make the passage take shape by stringing its words into sentences. If they can make the sentences, they can read the passage and then, and only then, begin to think about what

---

2. The familiarity of the words used in a passage can be estimated from any comprehensive word usage compilation *A Basic Vocabulary of Elementary School Children*, Henry D. Rinsland, 1945; *The Teacher's Word Book of 30,000 Words*, Edward L. Thorndike and Irving Lorge, 1944; *The Word Frequency Book*, John B. Carroll, Peter Davies, and Barry Richman, 1971; *The Educators' Word Frequency Guide*, Susan M. Zeno, Stephen H. Ivens, Robert T. Millard, and Raj Davvuri, 1995.

the passage has to say. Knowing the words and making the sentences sets the threshold for reading.

To Lexile a passage, we look up the occurrence frequency of each word. The Lexile Analyzer uses the average log word frequency and the logarithm of average sentence length. The final Lexile measure for the passage is a weighted sum of these two logarithms. Figure 33 shows how to Lexile a book. Figure 34 shows how to Lexile a reader. The coefficients in the formula are set to provide the most efficient balance between log word familiarity and log sentence length and to define a metric that reaches 1000 Lexiles from the books used in first grade at 200 Lexiles to the books used in 12th grade at 1200 Lexiles. The Lexile range of readability goes from (-200) to 1800. The equation is simple. Word familiarity and sentence length are all there is to it.

Step 1. Test the reader with L response-illustrtated Lexile calibrated items of

- average slice Lexile, H, and

- slice Lexile standard deviation, S.

Step 2. Count the reader's right answers for Score (R). The reader's Lexile measure is

$$\text{Reading Ability} = H + (180 + S^2/1040)\log[R/(L - R)] \quad (24)$$

The Lexile measure of a reader is equal to the Lexile level of a text for which the reader succeeds on 75% of the slices.

Figure 34. *How to Caculate a Lexile Reader Measure*

Lexile Relationships

Table 20 below and Table 21 on page 106 illustrate some useful Lexile relationships. When readers with a Lexile ability of 1000L are given a 1000L text, we expect them to experience a 75% success or comprehension rate (Stenner, 1992). If the same reader is given a 750L text, then we expect the rate to improve to 90%. If a text is at 500L, the rate should improve to 96%. The more readers' Lexile reading abilities surpass the Lexile readability of a text, the higher their expected success or comprehension rate. However, the more a text Lexile readability surpasses readers' Lexile reading abilities, the lower their expected rate.

Table 20. *Success Rates for Readers of Similar Ability with Texts of Different Readability*

Reader Ability	Text Readability	Text Titles	Expected Success
1000L	500L	*Are You There God? Its Me Margaret* (Blume)	96%
1000L	750L	*The Martian Chronicles* (Bradbury)	90%
1000L	1000L	*The Reader's Digest*	75%
1000L	1250L	*The Call of the Wild* (London)	50%
1000L	1500L	*On Equality Among Mankind* (Rousseau)	25%

Comprehension rates are relative. They are the results of Lexile differences between readers and texts. The 250L difference between a 750L text and a 1000L reader results in the same success rate as the 250L difference between a 1000L text and a 1250L reader. Each reader-text combina-

Table 21. *Success Rates for Readers of Different Ability with Texts of Similar Readability*

Reader Ability	*Sports Illustrated* Readability Lexile	Expected Success
500	1000	25%
750	1000	50%
1000	1000	75%
1250	1000	90%
1500	1000	96%

tion produces 90% reading success. Success rates are centered at 75% because readers forced to read at 50 percent success report frustration, whereas readers reading at 75% report comfort, confidence, and interest.[3]

All readers have their own range of reading comfort. As a result, there is a natural range of text readability that most motivates readers to improve their reading ability. Some readers are challenged by a success rate as low as 60%. Others find that burdensome. Once readers place themselves and their books in the Lexile Framework, they can discover what Lexile difference between their reading ability and text readability challenges them in the most productive way.

Book readability varies from page to page. Some books have a narrow range; that is, their passages cluster around a

---

3. Squires, Huitt, and Segars (1983) found that reading achievement for second graders peaked when their success rate reached 75%. A 75% rate is also supported by the findings of Crawford, King, Brophy, and Evertson (1975).

common level. As we read these books, the reading challenge stays level. There are no hills or valleys. Other books have a wide range of readability. There are easy passages and hard passages. These books can enable us to use the momentum that we gain from the easier passages to surmount the challenge of the harder ones. Overcoming this kind of resistance improves reading ability.

When we want to help students read, we can Lexile them and then offer them books with a readability that matches their reading ability. It is also helpful to know the book's passage difficulty variation. If we want our students to learn to read by reading, then we want to give them material that fascinates, motivates, absorbs, and also challenges them. We do that best by giving them books they want to read that are a little too hard for them, with passages that vary in passage difficulty. Then as they read along, they speed up and slow down. The speed-ups give them the energy and confidence needed to work through the slow-downs.

Using the Lexile Framework

Books are brought into the Lexile Framework by Lexiling the books. Tests are brought into the Framework by linking the target test scale (e.g., *SAT-9, TerraNova, Gates-MacGinitie*) with the Lexile Scale. Figure 34 illustrates how text and readers are measured on the same scale.

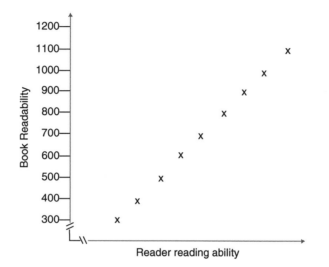

Figure 34. *Same Book Readability–Reader Reading Ability Scale*

To write a Lexile test item, we can use any natural piece of text. If we wish to write an item at 1000 Lexiles, we select books that contain passages at that level. We select a 1000-Lexile passage and add a relevant continuation sentence at the end with a crucial word missing. This is the "response illustration." Then we compose 4 one-word completions, all of which fit the sentence but only one of which makes sense in the logical context of the passage. Thus, the only technical concern is to ensure that all choices complete a perfectly good sentence, but that only one choice fits the passage. The correct answer for the response illustration in Figure 35 is C, repetition.

> You don't just establish a character once and let it go at that. Dominant impression, dominant attitude, dominant goal, all the rest — they must be brought forward over and over again; hammered home in scene after scene, so that the audience has no opportunity to forget them.
> **Use _____ for emphasis.**
>   A. humor
>   B. lighting
>   C. repetition
>   D. volume

Figure 35. *A 1000-Lexile Slice Test Item*

The aim of a Lexile item is to find out whether the student can read the passage well enough to complete the response-illustrated sentence with the word that fits the passage. Lexiled items like this are used to build theoretically parallel linking tests that are used in a common-person data collection design to link a target test scale with the Lexile Scale.

The Lexile Slice is a simple, easy-to-write item type. In practice, however, we may not even need the slice to determine how well a person reads. Instead, we may proceed as we do when we take a child's temperature. Because the Lexile Framework provides a ruler that measures readers and books on the same scale, we can estimate any person's reading ability by noting the Lexile level of the books they enjoy.

The 1-minute Self-report. When our child says, "I feel hot!" we infer that they have a fever. When a person says, "I like these books," and we know the books' Lexile levels, we can infer that the person reads at least that well (see Table 22).

The 3-minute Observation. To find out more about our child, we feel the forehead. The 3-minute way to measure a person's reading is to pick a book with a known Lexile level and ask the person to "Read me a page." If they read without hesitation, we know that they read at least that well. If they stumble, we pick an easier book. With two or three choices, we can locate the Lexile level at which the person is competent, just by having them read a few pages out loud. With a workbook of Lexile-calibrated passages, we can implement the 3-minute observation simply by opening the workbook and giving them successive passages to read.

Table 22. *Taking a Measure*

Method	Temperature	Reading
1-minute Self-report	I have a fever!	I like this book!
3-minute Observation	You feel hot!	Read this page.
15-minute Measurement	Your temperature is...	Your Lexile is...

The 15-minute Measurement. To find out more, we use a thermometer to take our child's temperature, perhaps several times. For reading, we give the person some Lexiled passages that end with an incomplete sentence. To measure reading ability, we find the level of Lexiled passages at which that person correctly recognizes what words are needed to replace the missing words 75% of the time.

The Lexile reading ruler connects reading, writing, speaking, and listening with books, manuals, memos, and instructions. This stable network of reproducible connections empowers a world of opportunities of the kind that the inch makes available to scientists, architects, carpenters and tailors (Luce & Tukey, 1964).

In school, we can measure which teaching method works best and manage our reading curriculae more efficiently and easily. In business, we can Lexile job materials and use the results to ensure that job and employee match. When a candidate applies for a position, we can determine ahead of time what level of reading ability is needed for the job and evaluate the applicant's reading ability by finding out what books they are reading and asking them to read a few sentences of job text out loud. This quick evaluation of an applicant's reading ability will show us whether the applicant is up to the job. When an applicant is not ready, we can counsel them, "You read at 800 Lexiles. The job you want requires 1000 Lexiles. To succeed at the job you want, you need to improve your reading 200 Lexiles. When you get your reading ability up to 1000, come back so that we can reconsider your application."

Lexile Perspectives

Job. Twenty-five thousand adults reported their jobs to the 1992 National Adult Literacy Survey (Campbell, Kirsch, & Kolstad, 1992; Kirsch, Yamamoto, Norris, Rock, et al., 2001). Their reading ability was also measured. Figure 36 summarizes the relationship between reading ability and employment (Wright & Stenner, 2000). In 1992, the average construction worker read at 1000 Lexiles. The average secretary read at 1200, the average teacher at 1400, and the average scientist at 1500.

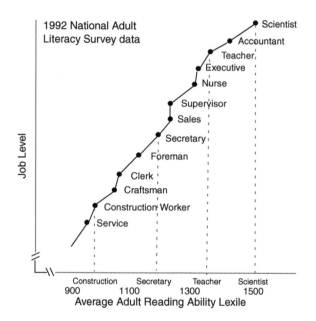

Figure 36. *Reading Ability Limits Employment*

When we can see so easily how much increasing our reading ability can improve our lives, we cannot help but be motivated to improve, especially when what we must do is so obvious. If we want to be a teacher at 1400 Lexiles but read at only 1000, it is clear that we have 400 Lexiles to grow to reach our goal. If we are serious about teaching, the Lexile Framework shows us exactly what to do. As soon as we can take 1400 Lexile books off the shelf and read them easily, we know we can read well enough to be a teacher. If we find that we are still at 1000 Lexiles, however, then we cannot avoid the fact that we are not ready to qualify for teaching, not yet, not until we teach ourselves how to read more difficult text.

School. Most children learn to read in school. Rasch analysis of the 1992 National Adult Reading Survey showed that there is a strong relationship between the last school grade completed and subsequent adult reading ability (Wright & Stenner, 2000). Figure 37 on page 114 shows that, on average, we are never more literate than the day we left school. The average 7th grade graduate reads at 800 Lexiles. The average high school graduate reads at 1150 Lexiles. College graduates can reach 1400 Lexiles. For many of us, the last grade of school we successfully complete defines our reading ability for the rest of our lives. Once we leave school — and no longer benefit from the reading challenges that school provides — we tend to stop increasing our reading ability. The overwhelming implication of Figure 37 is that, if we aspire to become a truly literate society, then we must maintain schooling for everyone and help everyone stay in school as long as possible.

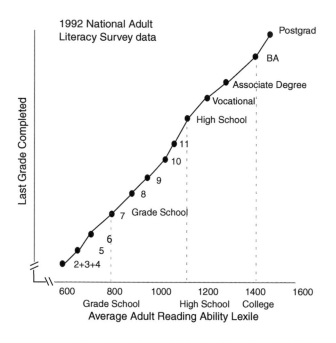

Figure 37. *Leaving School Limits Reading Ability*

Income. Reading ability also limits how much we can expect to earn. Figure 38 on page 115 shows the average incomes of readers in the 1992 National Adult Literacy Study at various Lexile reading abilities (Wright & Stenner, 2000). From 1000 to 1300 Lexiles, each reading ability increase of 150 Lexiles doubles our earning expectations. If we read at 1000 Lexiles and want to double our potential, then we have to improve our reading to 1150 Lexiles. When students can see the financial consequences of reading ability on an easy-to-understand scale that connects reading ability and income, then they have a persuasive reason to spend more time improving their reading abilities. The simple relationship shown in Figure 38 makes the road to riches obvious and explicit. No need to berate students, "Do your homework!" Instead, we can show them, "You want more money? You want to be a doctor? Here is the road. Learn to read better. It's up to you. We'll help you learn."

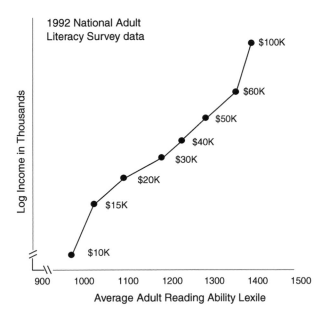

Figure 38. *Relationship between Reading Ability and Income Level*

Chapter 10   Summary

Data analysis requires the skillful interplay of theory and observation. In Chapters 1 through 6, we established how Rasch measurement provides the means for examining that relationship. Using Rasch measurement necessitates viewing the influence of persons and objects in light of their conjoint probability. Sir Isaac Newton and Charles Sanders Peirce were precursors to Rasch; in fact, Rasch's models utilize what has always been evident but not utilized in measurement until his formulations made such notions explicit. Guttman showed that conjoint order was paramount to scaling. Rasch measurement carries such conjoint probability to linear measures.

In Chapters 7 and 8, we presented two examples of Rasch measurement with data from a test and survey that were administered to clients of a metropolitan outpatient clinic. Chapter 7 illustrated a dichotomous response model and Chapter 8 a rating scale. We emphasized the important interplay between the scale intended for measurement and the empirical outcome. Well-reasoned intentions were stressed because they are fundamental. Evidence must be gathered and analyzed. These two tasks must interact successfully for measurement to emerge.

We illustrated strategies for observing the structure of data using examples from the measurement program that show the key elements in analysis: misfit and the study of residuals. Our goal was to show how Rasch measurement worked with a comprehensive software program to produce useful measurement.

Chapter 9 presented an example of using Rasch analysis to generate measures of readers and texts on the same scale.

We have illustrated the principles necessary for constructing a measure in sufficient detail to make them available to anyone interested in using them. The examples we used are from psychology and reading. The principles and methods can be fruitfully applied to any field of inquiry.

References

Bernoulli, J. (1713). *Ars conjectandi*. Basil: Thurnisiorum.

Bernstein, I. H., & Teng, G. (1989). Factoring items and factoring scales are different: Spurious evidence for multidimensionality due to item categorization. *Psychological Bulletin, 105*(3), 467–477.

Campbell, A., Kirsch, I. S., & Kolstad, A. (1992). *Assessing literacy: The framework for the National Adult Literacy Survey*. Washington, DC: National Center for Education Statistics, U.S. Department of Education.

Carroll, J. B., Davies, P., & Richmond, B. (1971). *The word frequency book*. Boston: Houghton Mifflin.

Chall, J. S. (1988). The beginning year. In B. L. Zakaluk & S. J. Samuels (Eds.), *Readability: Its past, present and future*. Newark, DE: International Reading Association.

Crawford, W. J., King, C. E., Brophy, J. E., & Evertson, C. M. (1975). *Error rates and question difficulty related to elementary children's learning*. Paper presented at the annual meeting of the American Educational Research Association, Washington, DC.

Davis, F. (1944). Fundamental factors of comprehension in reading. *Psychometrika, 9*, 185–197.

Educational Testing Service tests of applied literacy skills. (1990). New York: Simon & Schuster Workplace Resources.

Guttman, L. (1944). A basis for scaling qualitative data. *American Sociological Review, 9*, 139–150.

Jaeger, R. M. (1973). The national test equating study in reading: The anchor test study. *Measurement in education, 4,* 1–8.

Kirsch, I. S., & Jungeblut, A. (1986). *Literacy: Profiles of America's young adults final report* (ERIC No. ED275701). Princeton, NJ: National Assessment of Educational Progress, Educational Testing Service.

Kirsch, I., Yamamoto, K., Norris, N., Rock, D., Jungeblut, A., O'Reilly, P., et al. (2001). *Technical report and data file user's manual for the 1992 National Adult Literacy Survey.* Washington, DC: U.S. Department of Education: Office of Education Research and Improvement.

Linacre, J. M. (2000). Instantaneous measurement and diagnosis. *Popular Measurement, 1*(1), 55–59.

Linacre, J. M. (2003). WINSTEPS. Chicago: MESA Press.

Loret, P. G., Seder, A., Bianchini, J. C., & Vale, C. A. (1974). *Anchor test study final report: Project report and Vols. 1–30* (ERIC Nos. Ed 092 601–ED 092 631). Berkeley, CA: Educational Testing Service.

Lorge, I. (1939). Predicting reading difficulty of selections for school children. *Elementary English Review, 16,* 229–233.

Luce, R. D., & Tukey, J. W. (1964). Simultaneous conjoint measurement: A new type of fundamental measurement. *Journal of Mathematical Psychology, 1,* 1–27.

Miller, G. (1956). The magical number seven plus or minus two. *Psychological Review, 63,* 81–97.

Mitchell, J. V. (1985). *The ninth mental measurements yearbook.* Lincoln, NE: University of Nebraska Press.

Peirce, C. S. (1878). The probability of induction. *Popular Science Monthly, 12,* 705–718.

Piaget, J. (1950). *The psychology of intelligence* (M. Piercy & D. E. Berlyne, Trans.). London: Routledge and Kegan Paul.

Rasch, G. (1961). *On general laws and meaning of measurement.* Paper presented at the Proceedings of the Fourth Berkeley Symposium on Mathematical Statistics and Probability, Berkeley: University of California.

Rasch, G. (1980). *Probabilistic models for some intelligence and attainment tests.* Chicago: The University of Chicago Press (original work published in 1960).

Reder, S. (1996). Dimensionality and construct validity of the NALS assessment. In M. C. Smith (Ed.), *Literacy for the 21st century: Research, policy and practice.* Westport, CT: Greenwood Publishing Group.

Rentz, R. R., & Bashaw, W. L. (1975). *Equating reading tests with the Rasch model, Vol 1: Final Report & Vol. 1 & Vol 2: Technical reference tables. Final report of U.S. Department of Health, Education, and Welfare Contract OEC–072–5237* (ERIC Nos. ED 127 330 & ED 127 331). Athens: The University of Georgia.

Rentz, R. R., & Bashaw, W. L. (1977). The National reference scale for reading: An application of the Rasch model. *Journal of Educational Measurement, 14*, 161–179.

Rinsland, H. D. (1945). *A Basic vocabulary of elementary school children.* New York: Macmillan.

Rock, D. A., & Yamamoto, K. (1994). *Construct validity of the adult literacy subscales.* Princeton, NJ: Educational Testing Service.

Salganik, L. H., & Tal, J. (1989). *A Review and reanalysis of the ETS / NAEP young adult literacy survey.* Washington, DC: Pelavin Associates.

Sears, S. D. (1997). *A monetary history of Iraq and Iran.* Unpublished Ph.D. dissertation, University of Chicago.

Squires, D. A., Huitt, W. G., & Segars, J. K. (1983). *Effective schools and classrooms.* Alexandria, VA: Association for Supervisor and Curricular Development.

Stenner, A. J. (1992, April). *Meaning and method in reading comprehension.* Paper presented at the American Educational Research Association, Division, D, Rasch Special Interest Group, San Francisco, CA.

Stenner, A. J. (1996, February). *Measuring reading comprehension with the Lexile Framework.* Paper presented at the Fourth North American Conference on Adolescent/ Adult Literacy, Washington, DC.

Stenner, A. J. (1998, June). Writing Lexile items. Paper presented at the American Educational Research Association Annual Meeting, University of Chicago.

Stenner, A. J., & Wright, B. D. (2002). *Readability, reading ability, and comprehension.* Paper presented at the Association of Test Publishers Hall of Fame Induction for Benjamin D. Wright, San Diego.

Stone, M. H. (1998). Rating scale categories. *Popular Measurement, 1*(1), 61–65.

Stone, M. H. (2000, July 18). *Quality control in testing.* Paper presented at the Second International Congress on Licensure, Certification and Credentialing of Psychologists, Oslo, Norway.

Stone, M. H. (2002). *Knox cube test–revised.* Itasca, IL: Stoelting.

Thorndike, E. L., & Hagen, E. P. (1965). *Measurement and evaluation in psychology and education.* New York: Wiley.

Thorndike, E. L., & Lorge, I. (1944). *The teacher's word book of 30,000 words.* New York: Bureau of Publications, Teachers College.

Thurstone, L. L. (1946). Note on a reanalysis of Davis' Reading Tests. *Psychometrika, 11*(2), 185.

Wolpe, J., & Lang, P. J. (1964). A fear survey schedule for use in behaviour therapy. *Behaviour Research and Therapy, 2,* 27–30.

Woodcock, R. W. (1974). *Reading mastery tests*. Circle Pines, MN: American Guidance Service.

Wright, B. D., & Masters, G. N. (1981). *Rating scale analysis*. Chicago: MESA Press.

Wright, B. D., & Stenner, A. J. (1998, July 6). *Measuring reading*. Paper presented at the International Seminar on Developmental Assessment, Melbourne, Australia.

Wright, B. D., & Stenner, A. J. (2000). Lexile Perspectives. *Popular Measurement, 3*(1), 16-17.

Wright, B. D., & Stone, M. H. (1979). *Best test design*. Chicago: MESA Press.

Wright, B. D., & Stone, M. H. (1999). *Measurement essentials*. Wilmington, DE: Wide Range, Inc.

Zakaluk, B. L., & Samuels, S. J. (Eds.). (1988). *Readability: Its past, present and future*. Newark, DE: International Reading Association.

Zeno, S. M., Ivens, S. H., Millard, R. T., & Davvuri, R. (1995). *The educators word frequency guide*. New York: Touchstone Applied Science Associates, Inc.

Zwick, R. (1987). Assessing the dimensionality of the NAEP reading data. *Journal of Educational Measurement*, 24, 293–308.

Index

About the Authors

Ben Wright

Ben began his career in science as a research physicist, first at the Bell Telephone Laboratories with Nobel Laureate Charles H. Townes, investigating microwave absorption spectra of iodine monochloride, then at the University of Chicago with Nobel Laureate Robert S. Mulliken, investigating ultra violet absorption spectra of organic molecules. He became director of a group theater for young adults at the Gads Hill Center, Settlement House in Chicago. Concerns about care of emotionally disturbed children prompted him to investigate staff problems encountered in children's institutions. He designed research employing observation, life histories, interviewing, reports, test construction, and factor and variance analysis to assess treatment of childhood schizophrenia at the Orthogenic School of the University of Chicago under the direction of Bruno Bettleheim. Dissatisfied with measure reliability produced by conventional social science statistical procedures, he saw a need for a reliable new method. In 1960, he found a breakthrough in the work of Danish mathematician Georg Rasch. Ben founded the Measurement Evaluation and Statistical Analysis Laboratory (MESA) Psychometrics Laboratory at the University of Chicago. Developing and applying methods for constructing and verifying measures as director of MESA, he has taught hundreds of colleagues and students to understand and implement useful objective measurement and authored 12 books and leading statistics and measurement software packages.

Mark H. Stone

Currently, Mark is director of research and professor at the Adler of Professional Psychology of Chicago. He teaches courses in research methods, statistics and psychometrics, assessment of dementia, and other neuropsychological topics. He earned a graduate degree in music theory and musicology and doctorates in psychology. A licensed clinical psychologist, he maintains a consulting practice and donates time to a community health agency.

In 2001, he retired as the provost and vice president of the Adler School. He met Ben Wright in the 1970s when they worked as consultants. After Ben demonstrated the power of Rasch analysis, Mark suggested they write a book. *Best Test Design* (1979) was the result. They also co-authored *Measurement Essentials* (2000).

Mark is also a diplomate in Adlerian Psychology, a member of the National Register of Health Service Providers in psychology, and a nationally certified alcohol and drug counselor, and a certified supervisor.

A. Jackson Stenner

Cofounder and chairman of MetaMetrics Inc., Jack led a team in the development of a unifying metric for reading ability and text readability. This unique application of Rasch measurement resulted in the Lexile Framework for Reading. Jack has published more than 60 papers, monographs and books, primarily on measurement and evaluation methodology. He is now working on the Quantile Framework for Mathematics. He is president of the Institute for Objective Measurement and past president of the Professional Billiard Tour Association.